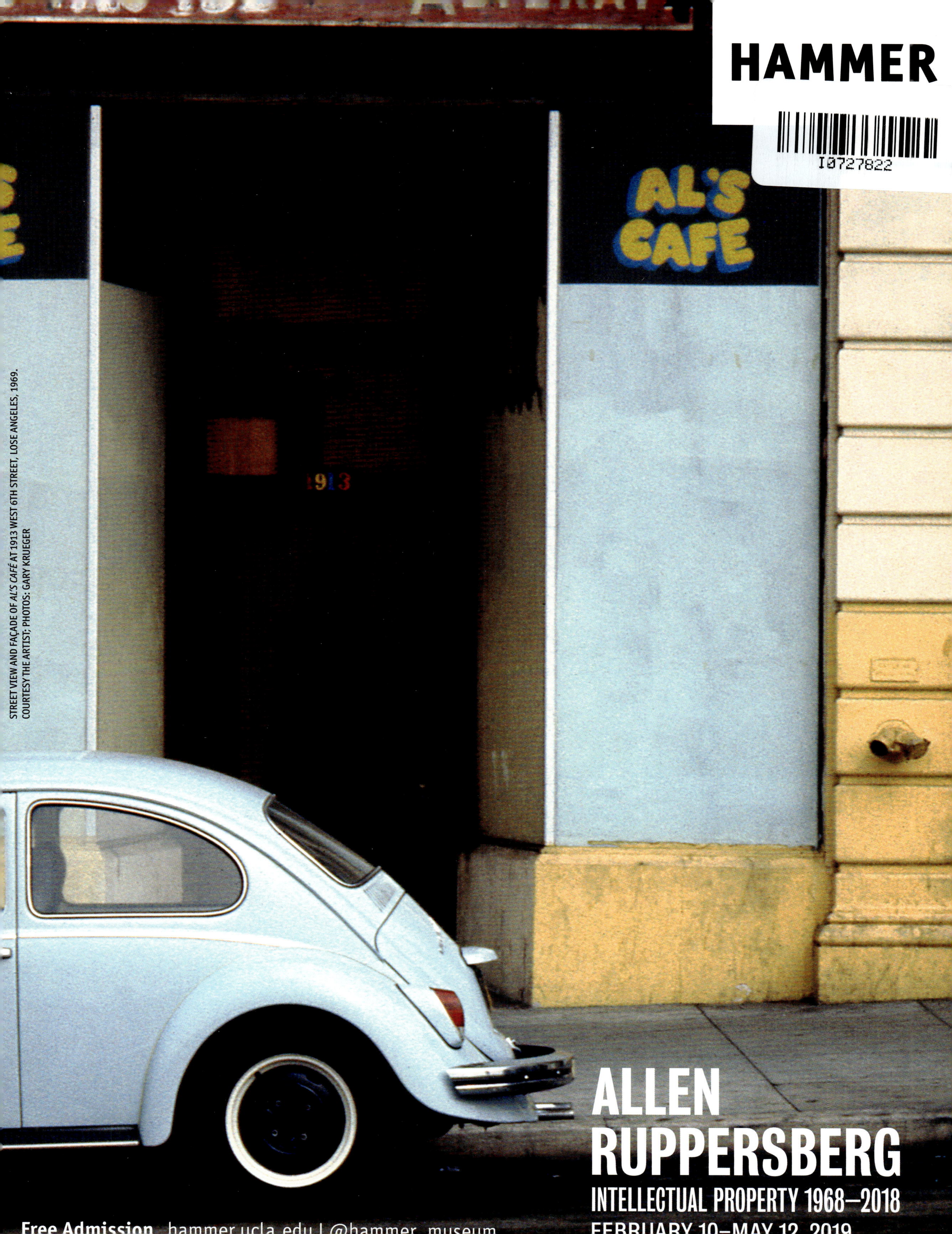

AL'S CAFE
1913
HAMMER
I0727822
ALLEN RUPPERSBERG
INTELLECTUAL PROPERTY 1968–2018
FEBRUARY 10–MAY 12, 2019
Free Admission   hammer.ucla.edu | @hammer_museum

Spring 2019

Words

Pictures

Front

Back

Opposite:
A. L. Steiner,
*ExxonMobil CEO:
Ending Oil Production
'Not Acceptable for
Humanity' (25 May 2016,
The Guardian)*, 2018
Courtesy the artist

Front cover:
David Benjamin Sherry,
Looking toward Valley
of the Gods, Bears Ears
National Monument,
Utah (detail), 2018
Courtesy the artist and
Salon 94

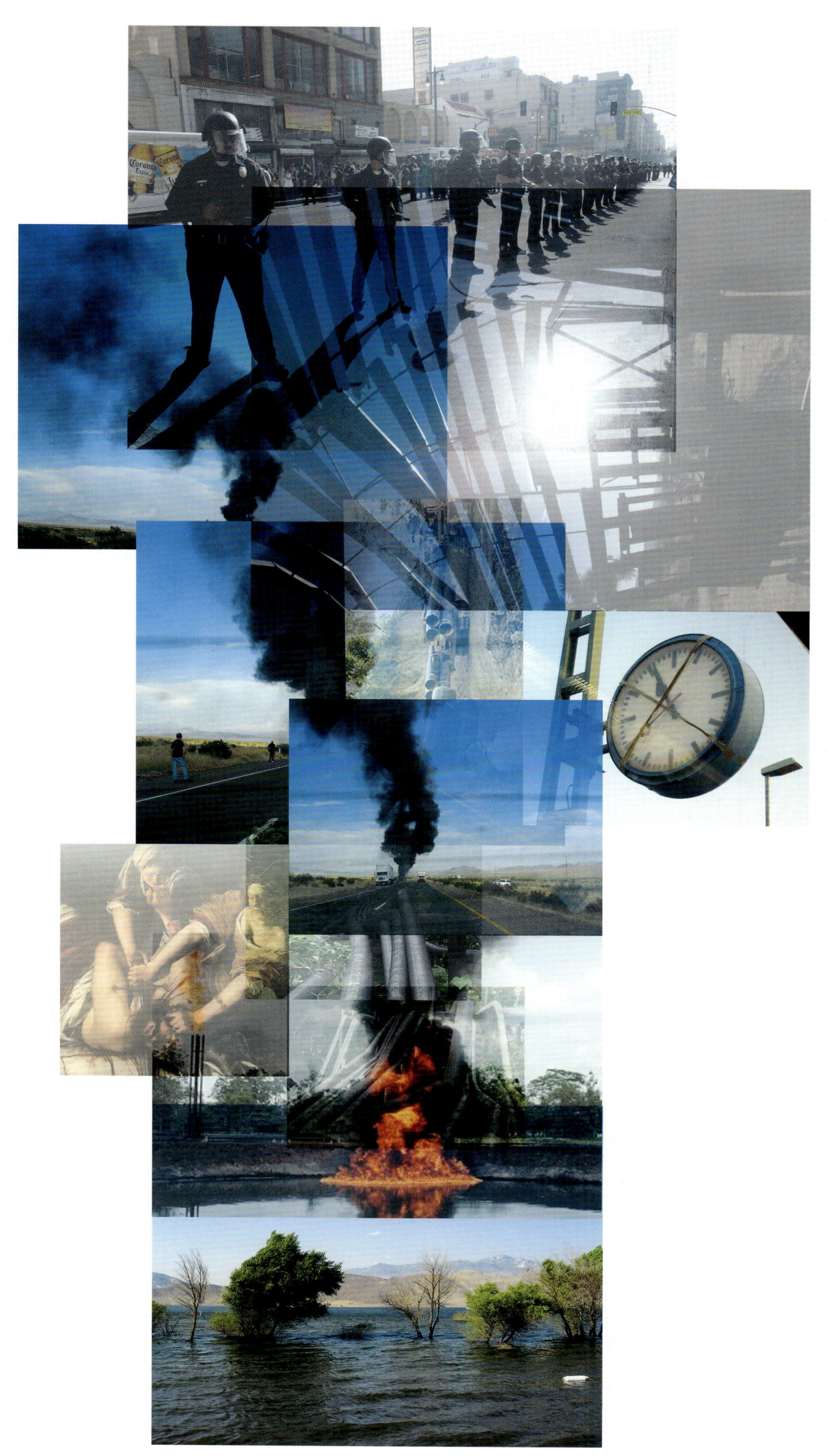

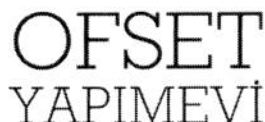

OFSET
YAPIMEVİ

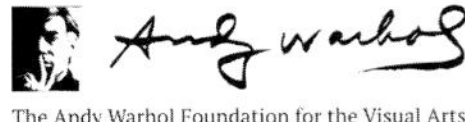

The Andy Warhol Foundation for the Visual Arts

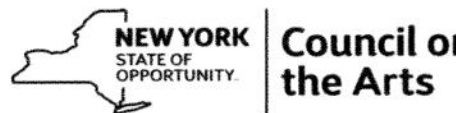

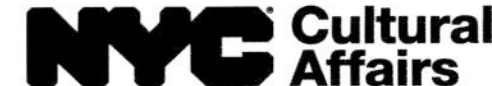

Statement of Ownership, Management, and Circulation (Required by 39 U.S.C. 3685). 1. Publication Title: Aperture; 2. Publication no.: 0003-6420; 3. Filing Date: October 1, 2018 4. Issue Frequency: Quarterly; 5. No. of Issues Published Annually: 4; 6. Annual Subscription Price: $75.00; 7. Complete Mailing Address of Known Office of Publication: Aperture Foundation, 547 West 27th Street, 4th Floor, New York, NY 10001-5511; Contact Person: Dana Triwush; Telephone: 212-946-7116; 8. Complete Mailing Address of Headquarters or General Business Office of Publisher: Aperture Foundation, 547 West 27th Street, 4th Floor, New York, NY 10001-5511; 9. Full Names and Complete Mailing Addresses of Publisher, Editor, and Managing Editor: Publisher: Dana Triwush, Aperture Foundation, 547 West 27th Street, 4th Floor, New York, NY 10001-5511; Editor: Michael Famighetti, Aperture Foundation, 547 West 27th Street, 4th Floor, New York, NY 10001-5511; Managing Editor: Brendan Embser, Aperture Foundation, 547 West 27th Street, 4th Floor, New York, NY 10001-5511; 10. Owner: Aperture Foundation, Inc., 547 West 27th Street, 4th Fl., New York, NY 10001; 11. Known Bondholders, Mortgagees, and Other Security Holders Owning or Holding 1 Percent or More of Total Amount of Bonds, Mortgages, or Other Securities: None; 12. Tax Status: The purpose, function, and nonprofit status of this organization and the exempt status for federal income tax purposes: Has Not Changed During Preceding 12 Months; 13. Publication Title: Aperture; 14. Issue Date for Circulation Data Below: Summer 2018 #231; 15. Extent and Nature of Circulation (Average No. Copies Each Issue During Preceding 12 Months; No. Copies of Single Issue Published Nearest to Filing Date): a. Total Number of Copies (Net press run): 16,318; 15,932; b. Paid Circulation: (1) Mailed Outside-County Paid Subscriptions Stated on PS Form 3541: 5,590; 5,471; (2) Mailed In-County Paid Subscriptions Stated on PS Form 3541: 5; 5; (3) Paid Distribution Outside the Mails Including Sales Through Dealers and Carriers, Street Vendors, Counter Sales, and Other Paid Distribution Outside USPS: 4,038; 3,615; (4) Paid Distribution by Other Classes of Mail Through the USPS: 35; 35; c. Total Paid Distribution: 9,668; 9,126; d. Free or Nominal Rate Distribution: (1) Free or Nominal Rate Outside-County Copies included on PS Form 3541: 342; 351; (2) Free or Nominal Rate In-County Copies Included on PS From 3541: 0; 0; (3) Free or Nominal Rate Copies Mailed at Other Classes Through the USPS: 103; 103; (4) Free or Nominal Rate Distribution Outside the Mail: 525; 600; e. Total Free or Nominal Rate Distribution: 969; 1,054; f. Total Distribution: 10,637; 10,180; g. Copies not Distributed: 5,681; 5,752; h. Total: 16,318; 15,932; i. Percent Paid: 90.9%; 89.6%; 16. Electronic Copy Circulation, a. Paid Electronic Copies: 1,141; 903; b. Total Paid Print Copies + Paid Electronic Copies: 10,809; 10,029; c. Total Print Distribution + Paid Electronic Copies: 11,778; 11,083; d. Percent Paid (Both Print & Electronic Copies); 91.8%; 90.5%; I certify that 50% of all my distributed copies (Electronic & Print) are paid above a nominal price. 17. Publication of Statement of Ownership: Will be printed in the Spring 2019 issue of this publication; 18. I certify that all information furnished on this form is true and complete. I understand that anyone who furnishes false or misleading information on this form or who omits material or information requested on the form may be subject to criminal sanctions (including fines and imprisonment) and/or civil sanctions (including civil penalties). Signature and Title of Editor, Publisher, Business Manager, or Owner: Dana Triwush, Publisher, October 1, 2018

aperture

The Magazine of Photography and Ideas

Editor
Michael Famighetti

Managing Editor
Brendan Embser

Assistant Editor
Annika Klein

Copy Editors
Clare Fentress, Donna Ghelerter

Senior Production Manager
True Sims

Production Managers
Nelson Chan, Bryan Krueger

Work Scholars
Charis Morgan, Ellen Pong, Kaija Xiao

Art Direction, Design & Typefaces
A2/SW/HK, London

Publisher
Dana Triwush
magazine@aperture.org

Director of Brand Partnerships
Isabelle McTwigan
212-946-7118
imctwigan@aperture.org

Advertising
Elizabeth Morina
917-691-2608
emorina@aperture.org

**Executive Director,
Aperture Foundation**
Chris Boot

Minor White, Editor (1952–1974)

Michael E. Hoffman, Publisher and Executive Director (1964–2001)

aperture.org

THOMAS STRUTH

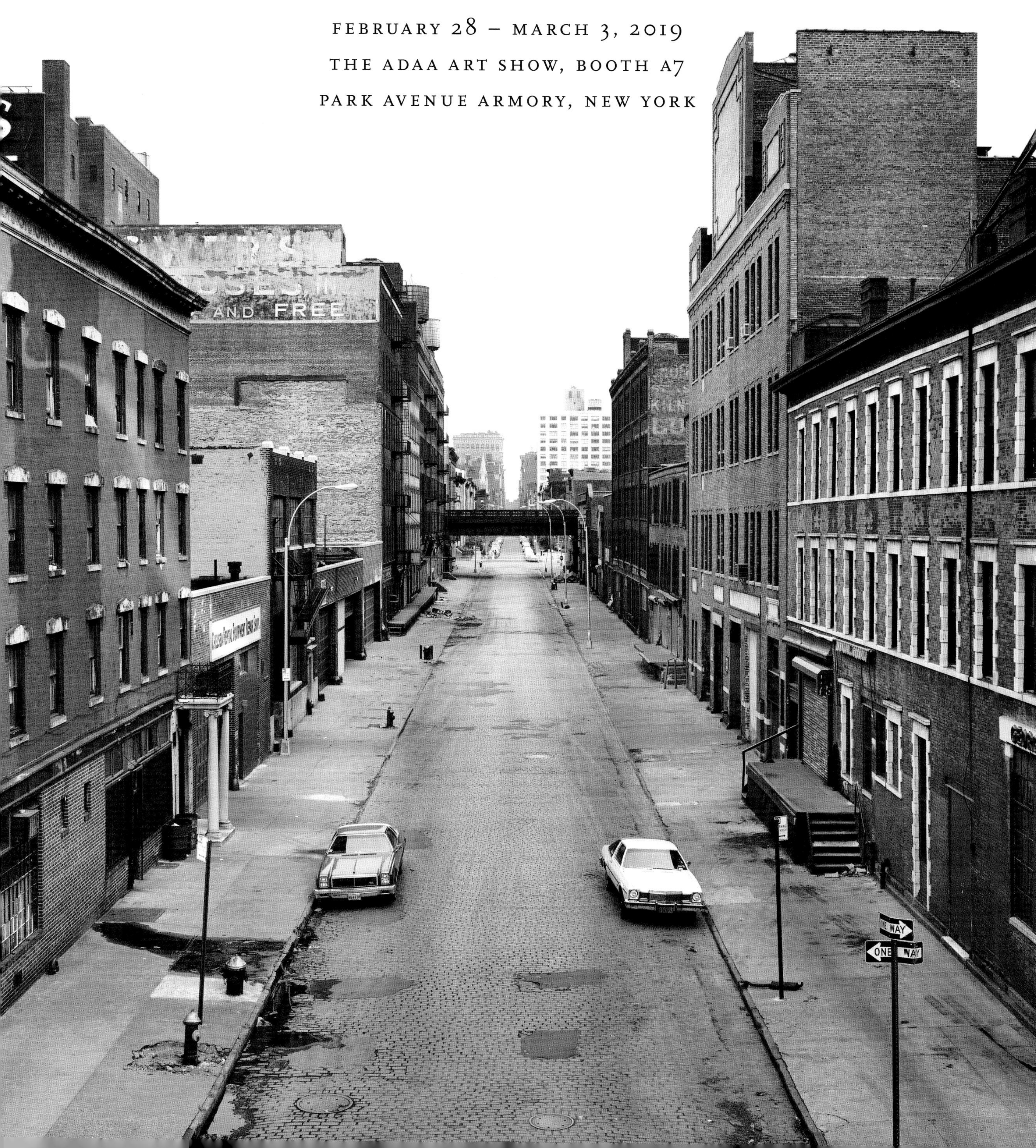

Agenda
Exhibitions to See

Encore

The past is prologue at the Getty, where, in the exhibition *Encore: Reenactment in Contemporary Photography*, seven artists assume the guises of figures from political, personal, or art-historical memory. Yasumasa Morimura inserts himself into a Manet. Samuel Fosso channels the spirits of civil rights–era icons and African independence leaders. Christina Fernandez re-creates the story of her mother's migration from Mexico to California, along the way adopting vernacular modes of twentieth-century photography. For artists from Yinka Shonibare MBE to Gillian Wearing, performance is central. As exhibition curator Arpad Kovacs notes, each "revises source materials from the past, either visual or written documents, or family narratives, and acts them out for the camera."

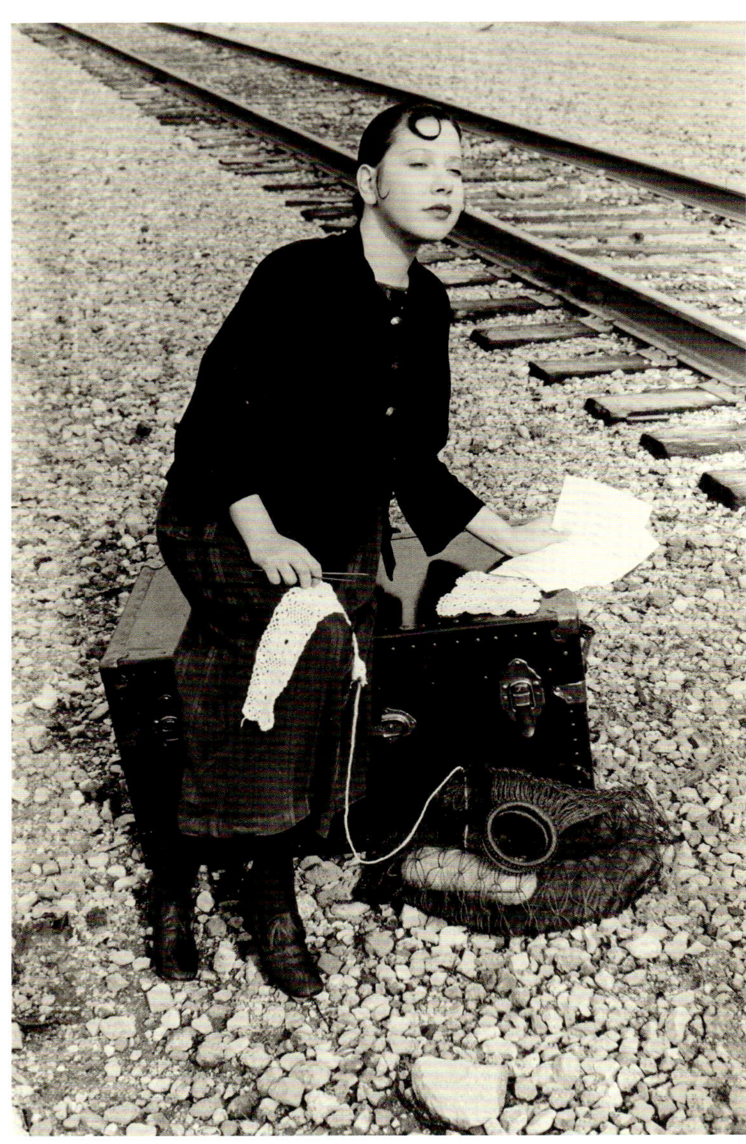

Christina Fernandez, *1927, Going Back to Morelia*, 1995
© the artist and courtesy the J. Paul Getty Museum, Los Angeles

Encore: Reenactment in Contemporary Photography at the Getty, Los Angeles, March 12–June 9, 2019

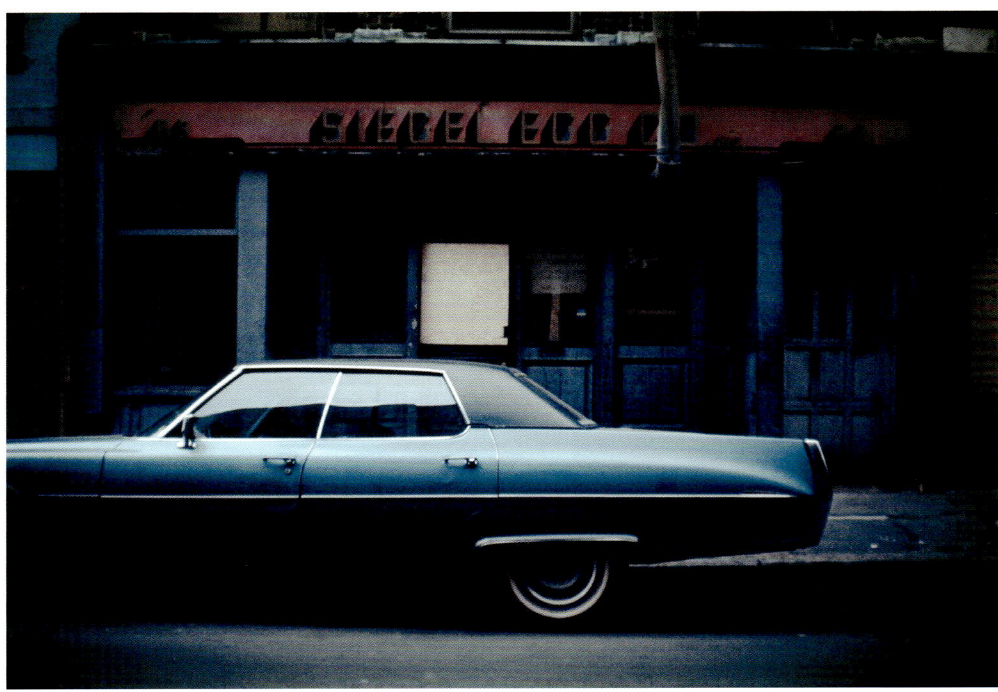

John Goodman, *Siegel Eggs / Boston*, **1973**
Courtesy the artist

John Goodman

As a young photographer juggling artistic work with commercial assignments, John Goodman always had color slide film loaded in a camera. Beginning in the 1970s and through the late 1980s, he photographed on the streets of Boston, finding fleeting moments of connection at diners, shops, and gas stations. He printed a few images, but packed most of his slides in a cabinet and only discovered them twenty-five years later, in 2009, when moving studios. Goodman, who would become known for his gritty series on New York's Times Square and Boston's Combat Zone, revisits this early work for the first time in *not recent color*, at the Addison Gallery of American Art. A former student of *Aperture* editor Minor White, Goodman made images that, for curator Allison Kemmerer, are "piercing and often moving views of what it is to be human."

John Goodman: *not recent color* at the Addison Gallery of American Art, Andover, Massachusetts, April 13–July 31, 2019

PHOTOGRAPHS
April 6 | New York | Live & Online

Including an Important Collection of Ruth Bernhard Photographs

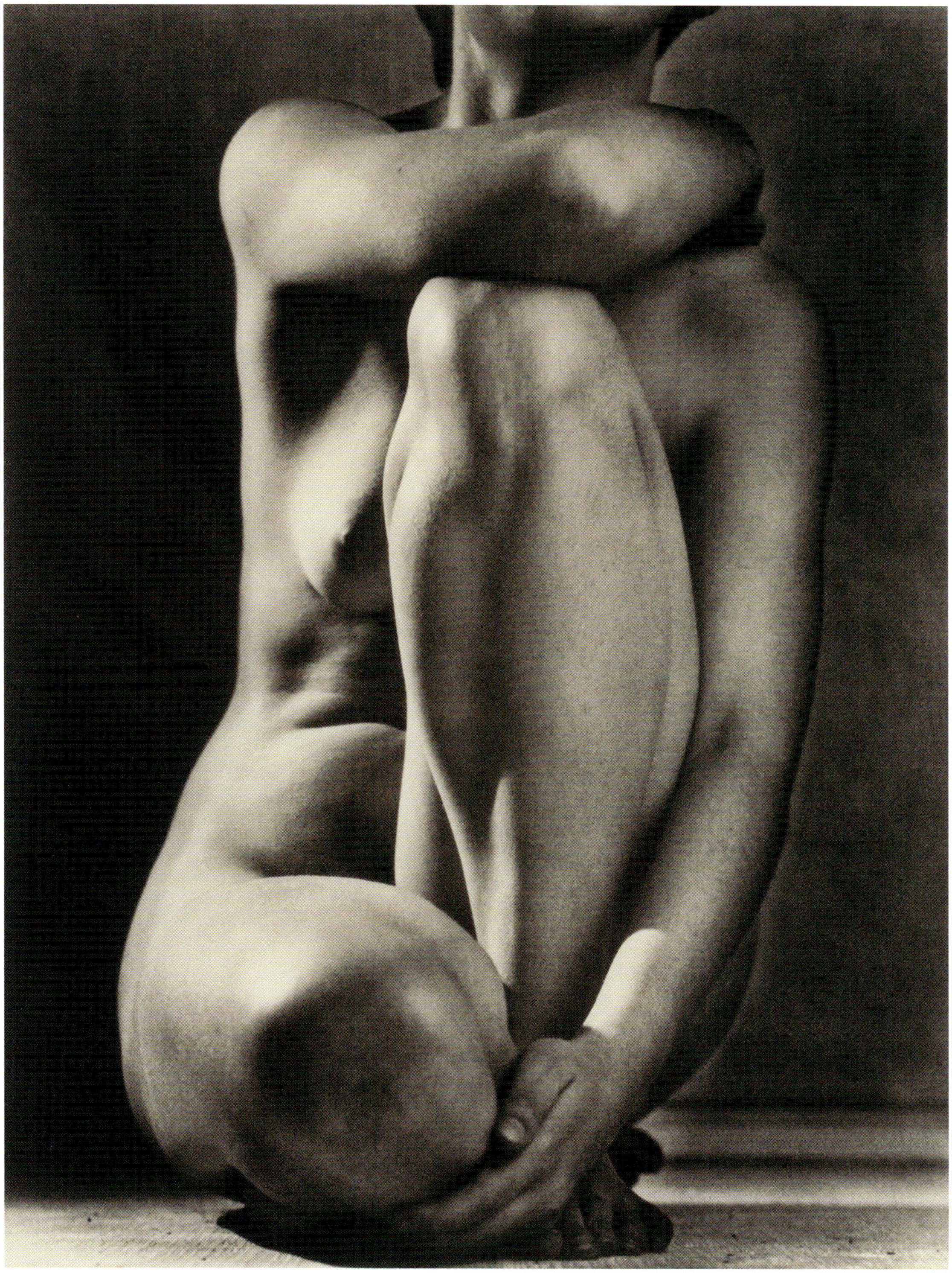

Ruth Bernhard
(American, 1905-2006)
Classic Torso, 1952
Gelatin silver, printed later

Inquiries:
Nigel Russell
212.486.3659 | NigelR@HA.com

VIEW | TRACK | BID
HA.com/5409

Always Accepting Consignments

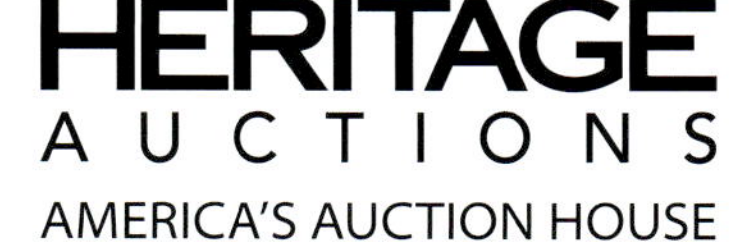

The Silvery Moon

"Almost from the moment photography was invented, photographers began dreaming about making photographs of the moon," says Diane Waggoner, curator of nineteenth-century photographs at the National Gallery of Art. "They wanted to bring together the power of the camera and telescope." Marking the fiftieth anniversary of the Apollo 11 moon landing, *By the Light of the Silvery Moon: A Century of Lunar Photographs from the 1850s to Apollo 11* spans technologies and styles to include a Walter de la Rue glass stereograph from the 1850s; richly detailed photogravures of early twentieth-century images by Charles le Morvan taken at the Paris Observatory; vernacular pictures of home televisions broadcasting the news, in July 1969, of the first moon landing; and Neil Armstrong's and Buzz Aldrin's own photographs of the moon's surface.

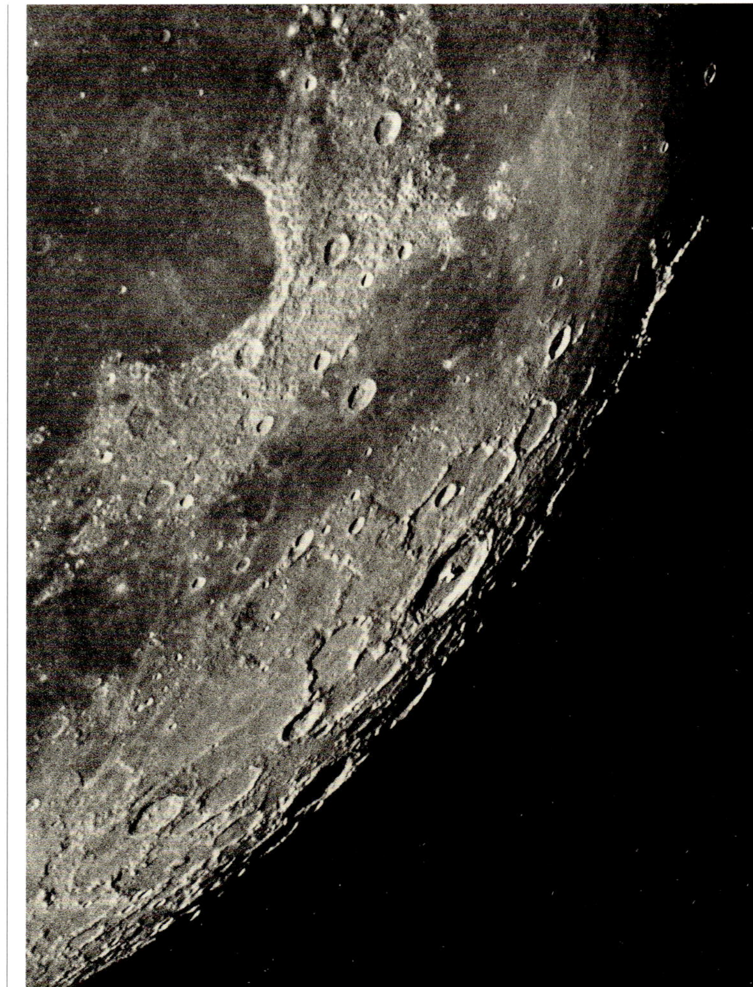

Charles le Morvan, *Carte photographique de la lune*, 1904
Courtesy the National Gallery of Art, Washington, D.C.

By the Light of the Silvery Moon: A Century of Lunar Photographs from the 1850s to Apollo 11 at the National Gallery of Art, Washington, D.C., April 28–October 14, 2019

Anne Collier

Are the women in Anne Collier's *Women Crying* (2016–ongoing) pictures sad, passionate, angry, or seductive? Like much of her work that considers the gendered history of the gaze, Collier's appropriated imagery invites endless interpretation. "She has a critical take on all the questions I'm interested in—representation in general, how gender roles and the female body are shown and discussed, and how our understanding as a society is shaped through photographic images," says Nadine Wietlisbach, director of Fotomuseum Winterthur, which will present selections from Collier's series *Women Crying* and *Women with Cameras* (2006–ongoing). Staging found pictures from the 1970s to the early 2000s in carefully composed still lifes, Collier dissects materials from mass media and pop culture to examine the mechanisms of photographic power.

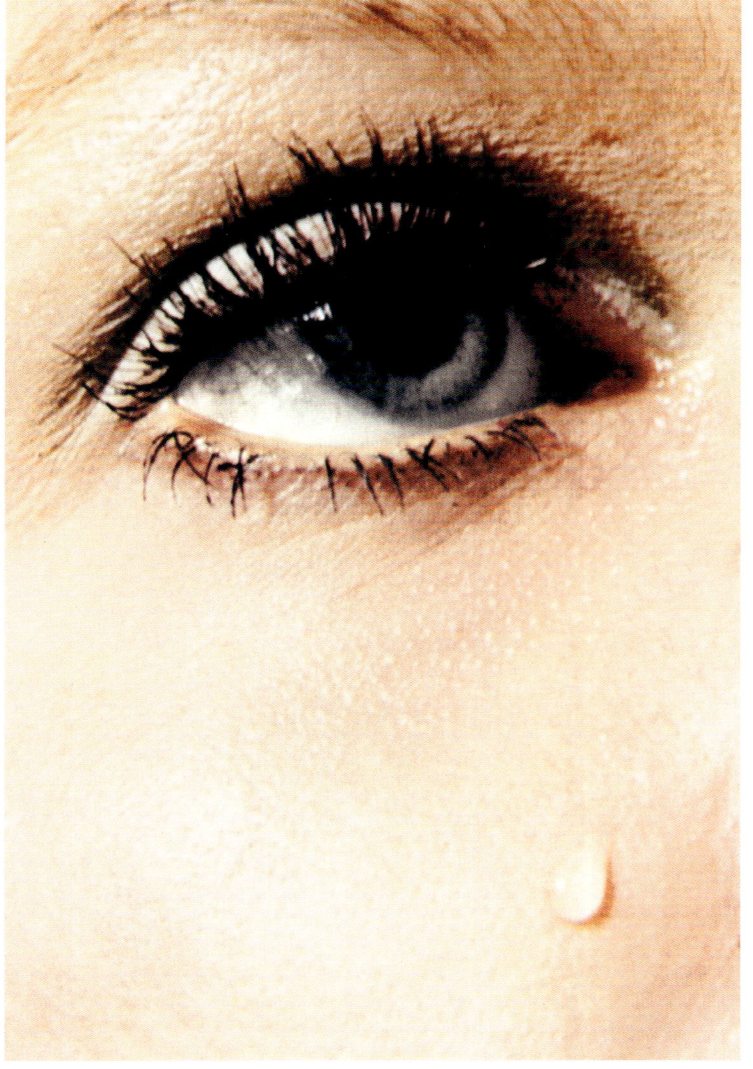

Anne Collier, *Woman Crying #7*, 2016, from the series *Women Crying*
© the artist and courtesy Anton Kern Gallery, New York

Anne Collier: Photographic at Fotomuseum Winterthur, Switzerland, February 23–May 26, 2019

THE INTERNATIONAL EXPOSITION OF
CONTEMPORARY AND MODERN ART

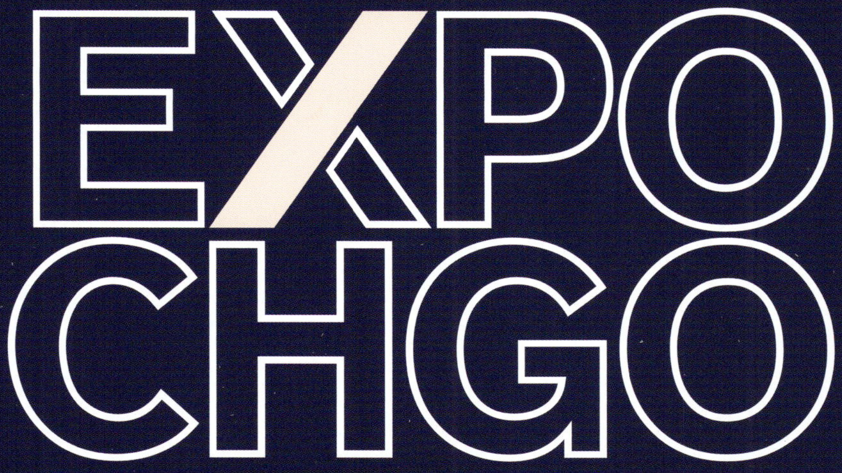

19–22 SEPTEMBER 2019

OPENING PREVIEW THURSDAY 19 SEPT

CHICAGO | NAVY PIER

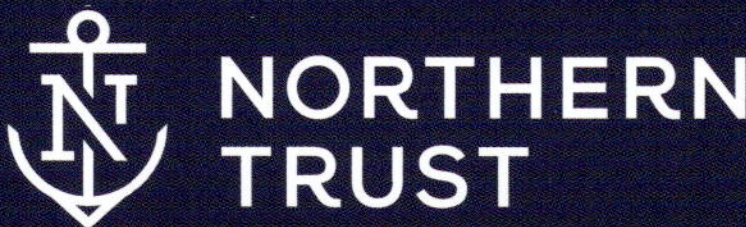

Presenting Sponsor

expochicago.com

IN ALIGNMENT WITH

CHICAGO
ARCHITECTURE
BIENNIAL

19 September 2019—
5 January 2020
Chicago Cultural Center & Citywide
chicagoarchitecturebiennial.org

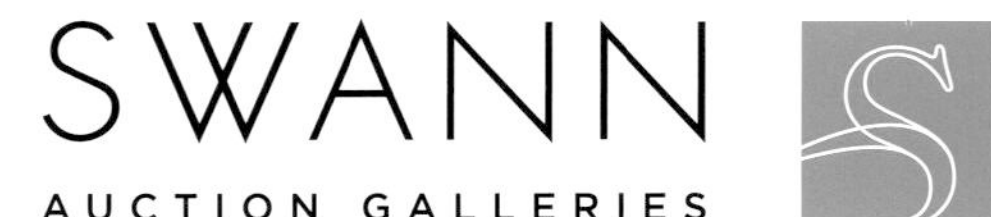

Peter Hujar, *David Wojnarowicz: Manhattan-Night (III)*, silver print, 1985. Estimate $15,000 to $25,000.

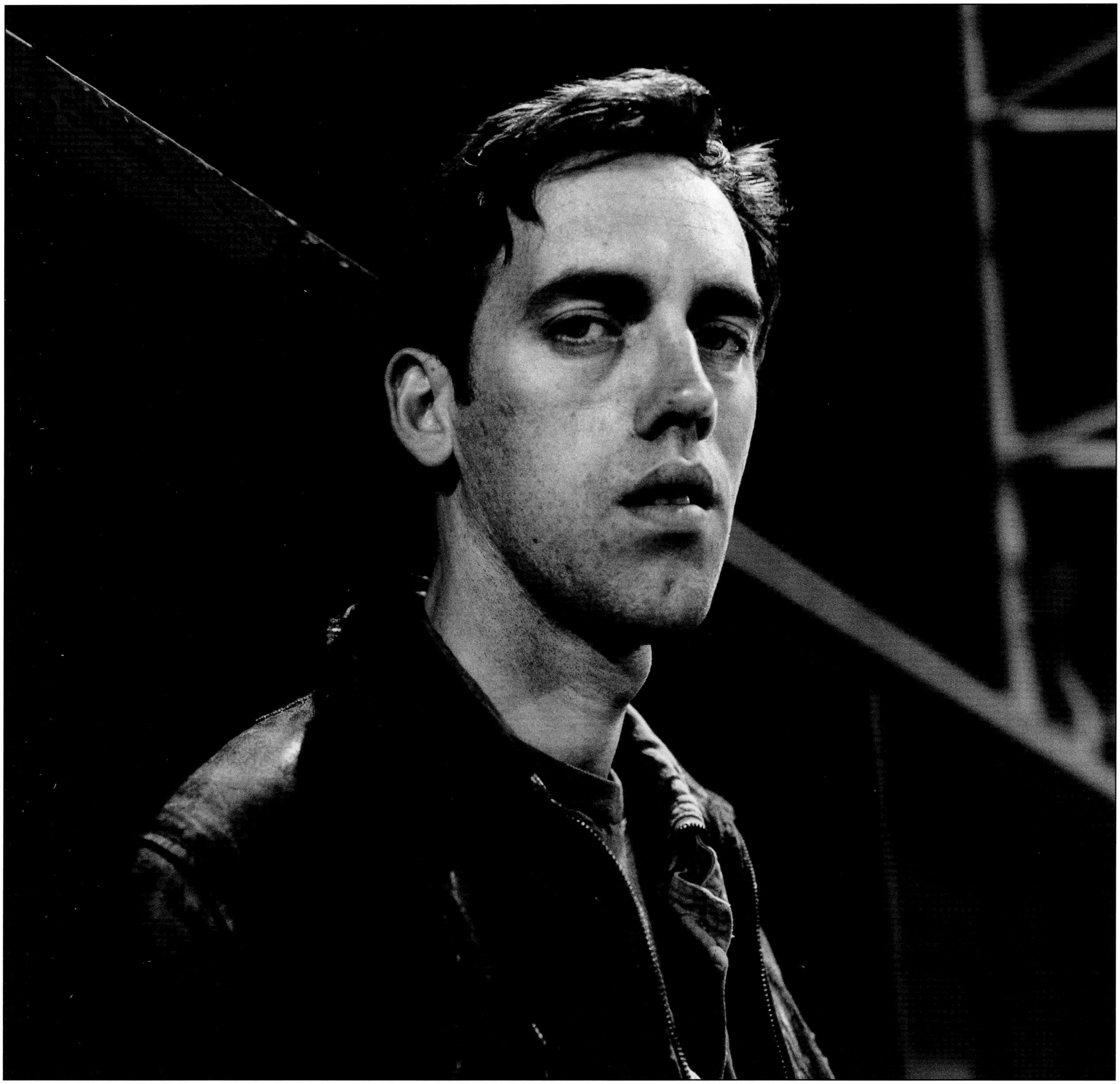

© Estate of Peter Hujar

The Pride Sale

June 20

Contact: Nicholas D. Lowry • pride@swanngalleries.com

104 East 25th Street New York, NY 10010 • tel 212 254 4710 • SWANNGALLERIES.COM

Backstory
Chuck Shacochis

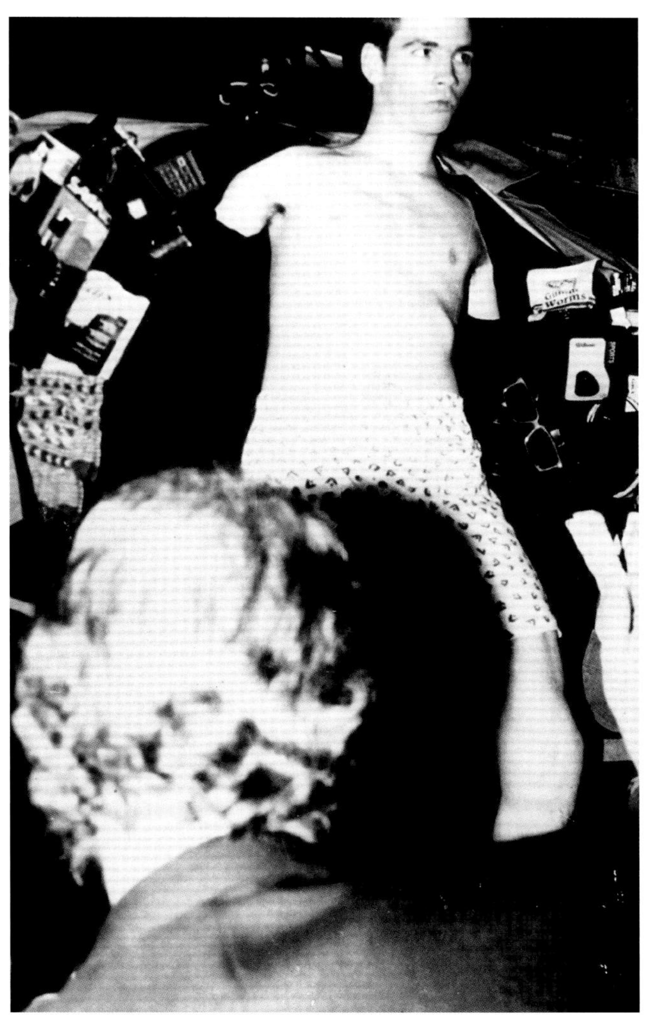

In John Waters's cult classic, a real photographer makes a fictional one famous
Lou Stoppard

More than twenty years have passed since the release of John Waters's film *Pecker* (1998), which tells the story of an aspiring photographer, played by Edward Furlong, whose deliciously gritty black-and-white images of the characters and happenings in Baltimore's Hampden neighborhood catch the eye of a New York gallerist. Much hullabaloo follows—critics rave over his debut show; an arts writer calls his friends and family "culturally challenged" in the newspaper; his grandmother Memama appears on the cover of *Artforum*; the Whitney Museum of American Art offers him a show, to be titled *A Peek at Pecker*; and, eventually, the great and good of the New York art world end up partying in Baltimore, where Pecker decides to stage his own show. The film remains a cult favorite and one in a long line of movies that buoy the cliché that all photographers are eccentric if lovable outsiders.

The images in the film were the work of then twenty-eight-year-old photographer Chuck Shacochis. Briefly, his life intersected with Pecker's, going from anonymity to the cusp of success. Before filming, he had served Waters various times at the camera shop where he worked in Baltimore. When Waters sought to book Matt Mahurin to capture Pecker's images—who, in a happy coincidence, Shacochis had on occasion traveled to New York to assist—Mahurin pointed out that, though he was unavailable, there was an ideal person right there in Baltimore.

"On set, John used to refer to me as the Lana Turner of the film—she got discovered from nowhere," says Shacochis. Underneath all the lampooning of the snobbism and vapidity of the art world, the film is an American-dream narrative, a story about how anyone can make it with sufficient talent and panache.

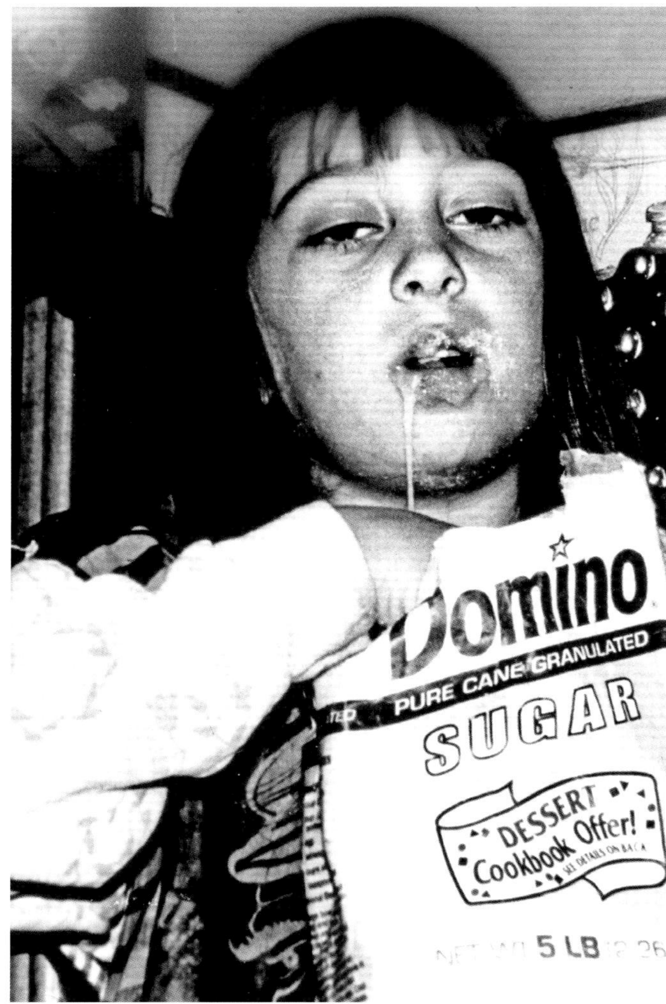

on set, rather than in advance, and the prints were being used immediately in the film. After a take, he would duck in and make an image, such as that of Pecker's kleptomaniac best friend, Matt, played by Brendan Sexton III, dancing on a Baltimore bar. (Matt has turned to stripping after being unable to shoplift unchallenged in the wake of Pecker's fame.)

After shooting, Shacochis's film would be rushed to a lab, so that it could be returned to him at the end of the day to be printed in his own darkroom through long nights and then presented to Waters the next morning. Sometimes he'd end up with only a couple of usable frames, such as when he was tasked with shooting two rats having sex; "Every day I'd be like, Please God, please God, let there be something here that can work." Waters wanted the photographs to look deliberately amateur, so Shacochis printed through a piece of glass that he covered in dirt and tea-bag stains. He made about thirty prints in total and was forced to turn all of the negatives over to the producers when filming ended.

Shacochis also found himself tasked with teaching the somewhat chaotic Furlong how to pass as a half-decent photographer: "I spent a while showing him how to use a camera, and how to move around a darkroom like you know what you're doing so it didn't look completely ridiculous—because photographers always complain about that kind of stuff in films about photography, and John was really concerned about it."

Like nearly all of Waters's movies, *Pecker* is a love letter to Baltimore. "It's an amazing place," says Shacochis, "but it's also a giant pain in the ass because there is so much possibility here that just never gets realized."

Pecker's own nonchalance in the face of all the opportunities promised by New York—"God, I don't wanna be in *Vogue*," he says at one point—is ironic, given that Shacochis was unable to turn the images into a viable career, despite the strength of the work and a succession of calls after the film's release.

Plans to move to New York, including one to enroll at Pratt Institute, never worked out, often due to family issues or financial restraints. "I had people saying, 'What's wrong with you? Why haven't you turned this into insane fame and fortune?' That happened a lot, and, for a while, I got bitter about it," he recalls. Shacochis still lives in Baltimore and continues to make portraits in the Hampden area—his work is tender and less sensationalist than the Pecker images. His pictures, he says, are images of "really quiet, nondescript moments."

Filming was stressful. While various photographers have produced images credited to fictional photographers in films—Don McCullin for Michelangelo Antonioni's 1966 *Blow-Up* and Steve Pyke and Mick Lindberg for Mike Nichols's 2004 *Closer*—the *Pecker* setup was unusual, as Shacochis was taking photographs

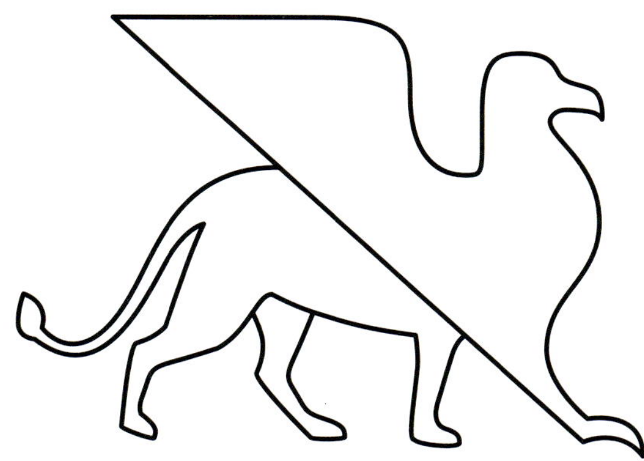

Griffin Editions

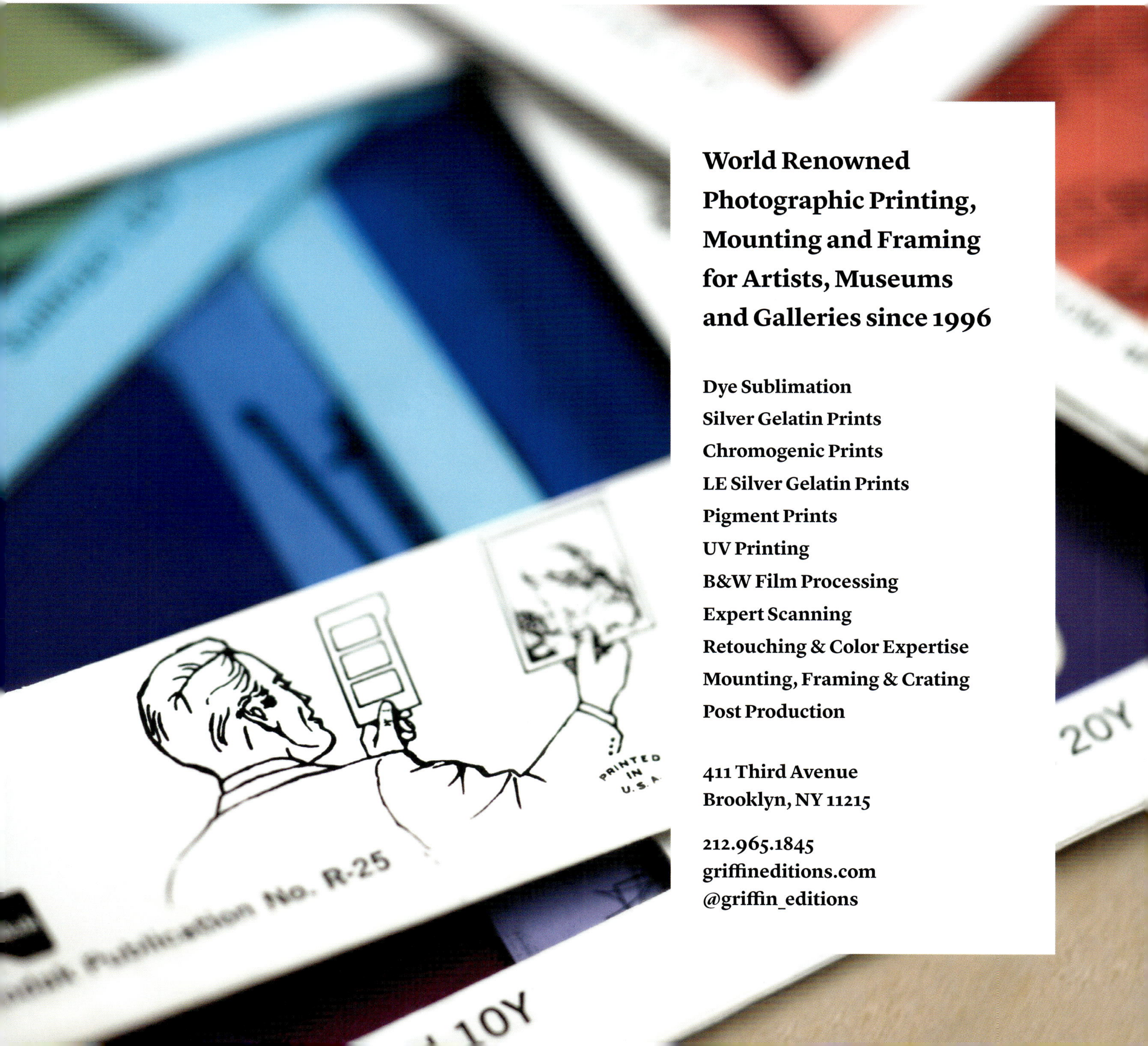

**World Renowned
Photographic Printing,
Mounting and Framing
for Artists, Museums
and Galleries since 1996**

Dye Sublimation
Silver Gelatin Prints
Chromogenic Prints
LE Silver Gelatin Prints
Pigment Prints
UV Printing
B&W Film Processing
Expert Scanning
Retouching & Color Expertise
Mounting, Framing & Crating
Post Production

**411 Third Avenue
Brooklyn, NY 11215**

**212.965.1845
griffineditions.com
@griffin_editions**

Aperture Limited-Edition Photographs

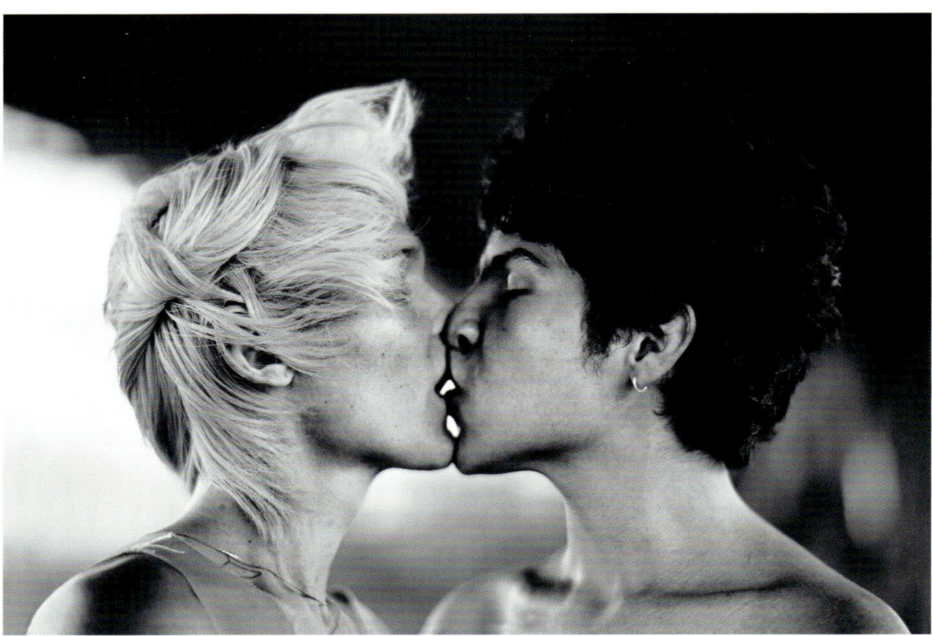

Aperture is pleased to present a series of limited-edition prints and portfolios from artists featured in the magazine. Proceeds from the sale of these editions support the artists as well as Aperture Foundation's publishing and public programs.

Ethan James Green
Gogo and Ser, 2015
Edition of 10. Featured in issue #229, "Future Gender."

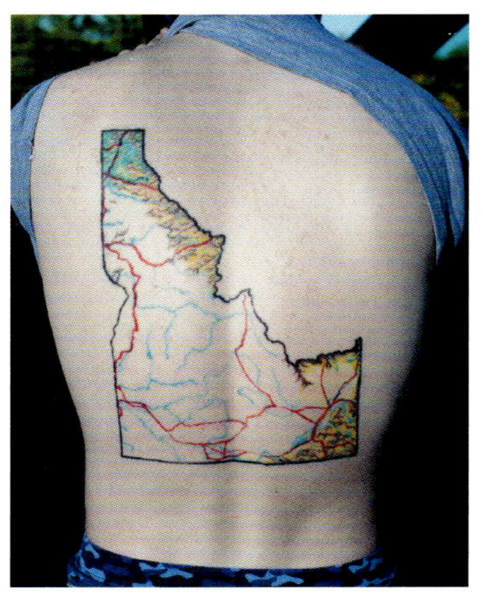

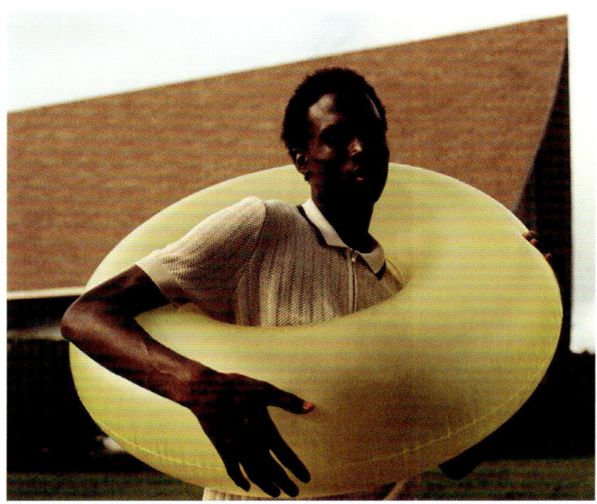

Carolyn Drake
Kyle in Target parking lot, Twin Falls, Idaho, 2016
Edition of 10. Featured in issue #226, "American Destiny."

Durimel
Bigger then, Bigger Glenn, 2017
Edition of 15. Featured in issue #228, "Elements of Style."

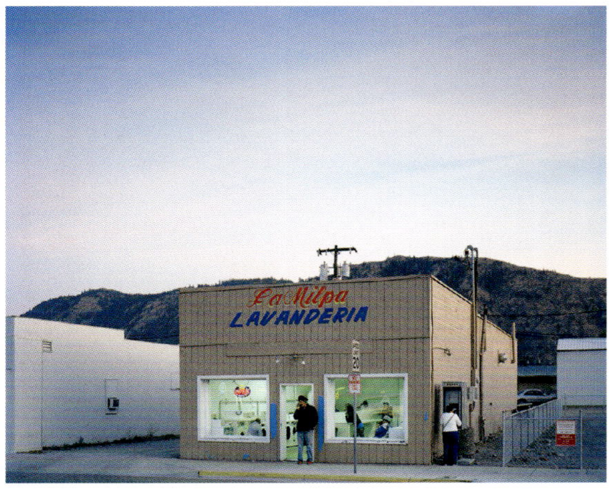

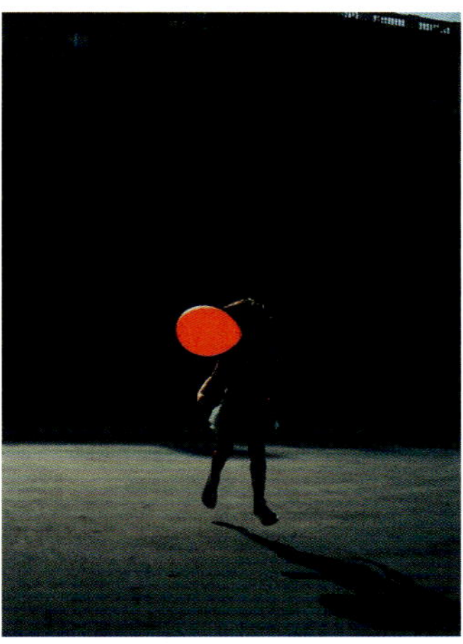

Kathya Maria Landeros
Main Street laundromat, Methow Valley, Washington, 2012
Edition of 10. Featured in issue #226, "American Destiny."

Christopher Anderson
Pia with balloon in Gràcia, Barcelona, 2016
Edition of 25. Featured in issue #233, "Family."

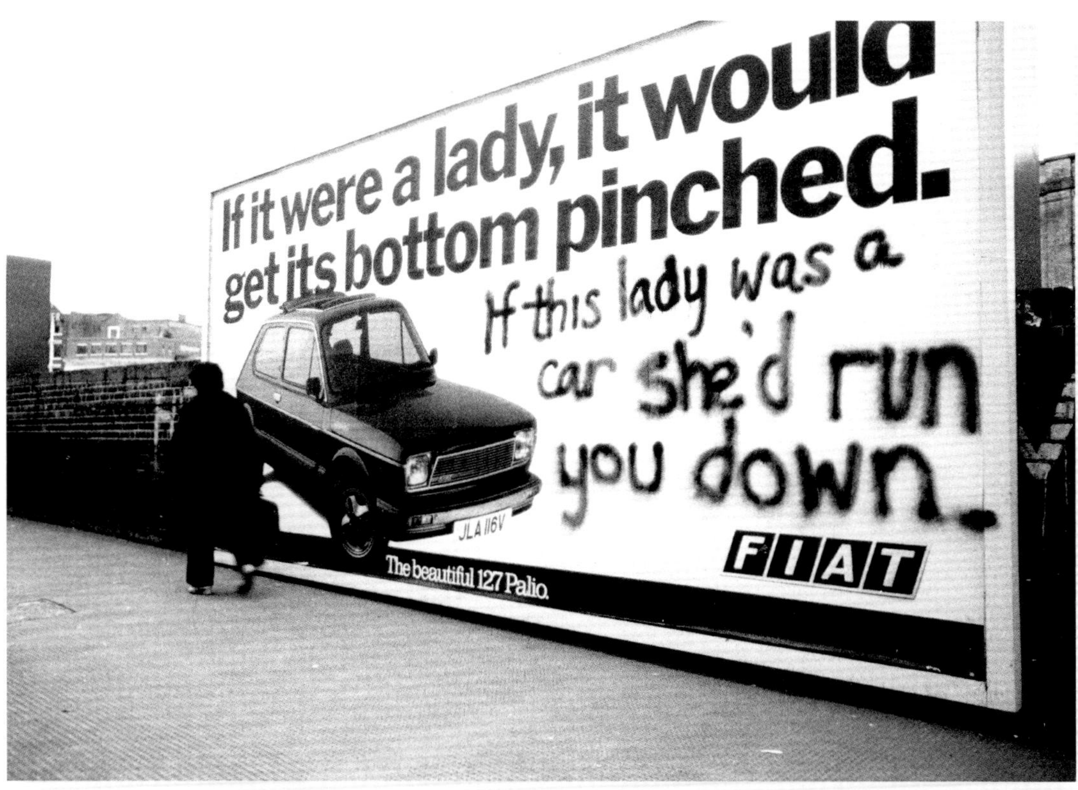

Signs of the times in Thatcher's Britain
Laura Guy

Sometimes rude, often crude, the graffiti that Jill Posener began photographing in London in the 1970s and collected in *Spray It Loud*, a slim photobook published in 1982, reflects the signature struggles of the British left at the time. The book's typology of delinquent typography is comprised of feminist tags and antiracist slogans, calls for social housing and for nuclear disarmament. The daubed demands that Posener encountered as she roamed the city with her A-to-Z street atlas appear against a backdrop of urban decay and social deprivation—the context that secured the Conservative Party its mandate to govern in 1979 with a pernicious ideology of free-market capitalism and nationalistic rhetoric. The graffiti, Posener writes in the book's introduction, reminds her "that there is resistance and rebellion."

Spray It Loud is a record of an ephemeral sign making that marks a place and a time. It's a manual, too, one dedicated to "the spray painters who take all the risks." A card featuring tips for dealing with an arrest is reprinted alongside the number for a twenty-four-hour emergency lawyer. Posener's images of words are also images of actions. Reproduction in print extends the readership that these slogans command. The book is a protest inasmuch as it augments the resistance that Posener describes.

Posener, who was the first lesbian playwright to work with the Gay Sweatshop Theatre Company, a collective founded in 1974, was committed to many of the causes represented in *Spray It Loud*. Her best-known images underscore the intrusion of feminist politics into the public sphere. This is a politics that forcibly takes its own permissions. A billboard advertising a new Fiat model reads: IF IT WERE A LADY, IT WOULD GET ITS BOTTOM PINCHED. To which someone scrawls the riposte, IF THIS LADY WAS A CAR SHE'D RUN YOU DOWN. The Fiat image was reproduced as a postcard that circulated in such high numbers that a reporter for *Woman of Power* magazine said Posener's pictures stood for a "populist expression of feminism." Certainly the caustic wit of the one-liners runs counter to the idea of feminists as pious politicos. WE CAN IMPROVE YOUR NIGHTLIFE, reads an advertisement for Rest Assured beds. JOIN LESBIANS UNITED, a vandal replies.

Revisiting *Spray It Loud*, I think of the refrain of a joke that might be repeated at different moments of feminist organizing, its punch line resonating through the entertaining placards carried on women's marches internationally, the media-savvy actions against cuts to services for domestic-abuse survivors in the U.K., and the epithets dashed off in toilet cubicles before anti-abortion laws were repealed recently in the Republic of Ireland. At this moment when the veracity of feminist statements is so blatantly governed by legislative denials, *Spray It Loud* speaks of the politically necessary task of seeking legibility on our terms, against their walls.

Jill Posener, Farringdon, London, 1979, from *Spray It Loud* (London: Routledge & Kegan Paul, 1982)

Laura Guy is an Early Career Academic Fellow in Art History at Newcastle University, U.K.

Curriculum
A List of Favorite Anythings
By Mark Steinmetz

"I love the South for its warmth and chaos," says photographer Mark Steinmetz, who was born in New York but now lives in Athens, Georgia. Steinmetz worked with Garry Winogrand in the 1980s and has photographed throughout the American South, as well as in Paris and various Italian cities, making silvery black-and-white images that conjure a mood of chance. For his most recent series, *ATL / Terminus* (2018), Steinmetz turned his lens on Hartsfield-Jackson Atlanta International Airport—the busiest airport in the world—and its surroundings. In his pictures of this bustling hub, he gives form to the poetic ambiance of the airport as a space of coming and going, conveying the open-endedness and quiet complexity of daily life.

Late work of Garry Winogrand

Garry Winogrand's photographs from the last years of his life—he died in 1984—strike me as being among the most interesting photographs ever made. The form becomes more unfamiliar and more scattered (shattered?); Winogrand was always loose, but these are looser yet. The mood hanging in the air over early 1980s Los Angeles is elegiac. Perhaps on some unconscious level he knew there was a cancer raging in his body.

The Century of Titian: The Golden Age of Venetian Painting, 1993

The best art exhibition I have ever seen is *The Century of Titian: The Golden Age of Venetian Painting*, held at the Grand Palais, in Paris, in 1993. Giovanni Bellini and Giorgione painted the most sublime images of human beings that I know of. A nocturnal scene by Jacopo Tintoretto of Christ praying in the olive grove was made with impressionistic strokes and fauvist colors hundreds of years before those movements took place in France.

Vic Chesnutt and Benjamin Smoke

Vic Chesnutt was raised in Zebulon, Georgia, and began to write songs at the age of five. A car accident at eighteen left him partially paralyzed. Most of his songwriting took place in Athens, Georgia (my adopted hometown). In "Panic Pure," Chesnutt sings: "My earliest memory is of holding up a sparkler / High up to the darkest sky / Some Fourth of July spectacular / I shook it with an urgency / I'll never be able to repeat." Listen to his 1992 album *West of Rome*. Benjamin Smoke, who worked in nearby Atlanta, put out brilliant albums at about the same time as Chesnutt. He might be more searchable as the Opal Foxx Quartet. Both died too young.

John Szarkowski, "A Different Kind of Art," 1975

This essay by John Szarkowski, published in the *New York Times* on April 13, 1975, argues with humor and grace that photography is unlike the other, more synthetic arts. He hits the nail on the head: the photographer's "entire effort is directed toward the problem of defining precisely what the subject is. This is meant not poetically but literally." I read this article every few years—it helps return me to clarity.

Rumi, Hafiz, and Basho

The thirteenth-century Sufi poet Jalal al-Din Rumi is among the most popular poets in America largely due to the renderings of his work by Coleman Barks, also of Athens, Georgia. I also love the ecstatic and mystical poetry of the fourteenth-century Sufi poet Hafiz—read *The Gift* (1999), translated by Daniel Ladinsky. These Persian poems are full of warmth and color. For something a little more cool and subdued, I might turn to Japan's Basho, seventeenth-century master of the haiku.

Fra Angelico

This is really what we all want: to have an angel come down and tell us God has something in mind for us. One of my favorite versions of the Annunciation is the fifteenth-century altarpiece by Fra Angelico in Cortona, Italy, a Tuscan hill town where I spent a few seasons. The current setting of this golden masterpiece, Cortona's Museo Diocesano, is quiet and intimate—there's nothing else around to crowd the altarpiece.

Orson Welles in his late movies

Andrew Sarris wrote of Orson Welles: "The conventional American diagnosis of his career is decline, pure and simple, but decline is never pure and never simple." In his late work—from *Touch of Evil* (1958), in which Marlene Dietrich tells Welles's character, "I didn't recognize you. You should lay off those candy bars…. You're a mess, honey," to *Chimes at Midnight* (1965), in which Welles's role is based on Shakespeare's Falstaff and Jeanne Moreau costars—Welles photographs his bloated, aging body and face in raking light and in wide-angle close-ups from high and low vantage points. Welles was a master of cinema, as well as of the fake nose.

Old stone tools

Pretty much all of the stone tools ever made—arrowheads, axes, choppers—are still with us today, they're just buried deep in the ground somewhere. I have a small collection of chopping tools that were possibly used to remove marrow from bone. Early hominids made them by striking one side of a cobble with another rock to create a sharp edge. After perhaps a million years of working this way, our ancestors came to realize that they could turn the stone over to make a sharper, double-sided edge. These sculptural choppers ground me to the earth and connect me to the distant past.

Pots by Michael Simon

Michael Simon is a potter who lives in Athens, Georgia (another artist working in my hometown). He was a student of the American Warren MacKenzie, who studied under the great potter Bernard Leach, who brought the Taoist stances of Japanese ceramicists to England. Simple and utilitarian, a beautiful and well-balanced cup is a joy.

Coins of Ancient Greece

Around 400 BCE in places like Syracuse, in Sicily, and Bruttium (now Calabria), in southern Italy, coins were made with astonishing artistry. Fanciful animals and striding nudes, like little Matisse drawings, were stamped into silver and gold and used as currency. I often hear people today make the assertion that we are the most evolved generation yet, and, in some ways, maybe we are, but not in all ways.

Late work of Eugène Atget

It cheers me to think of the elderly Eugène Atget getting up before dawn in mid-1920s Paris to take the train to the abandoned Parc de Sceaux, hauling his wooden tripod and heavy glass plates. There was no possibility for profit or acclaim (as the audience for his work hadn't arrived yet). He spent his early mornings communing with the gods in the bright, hazy light.

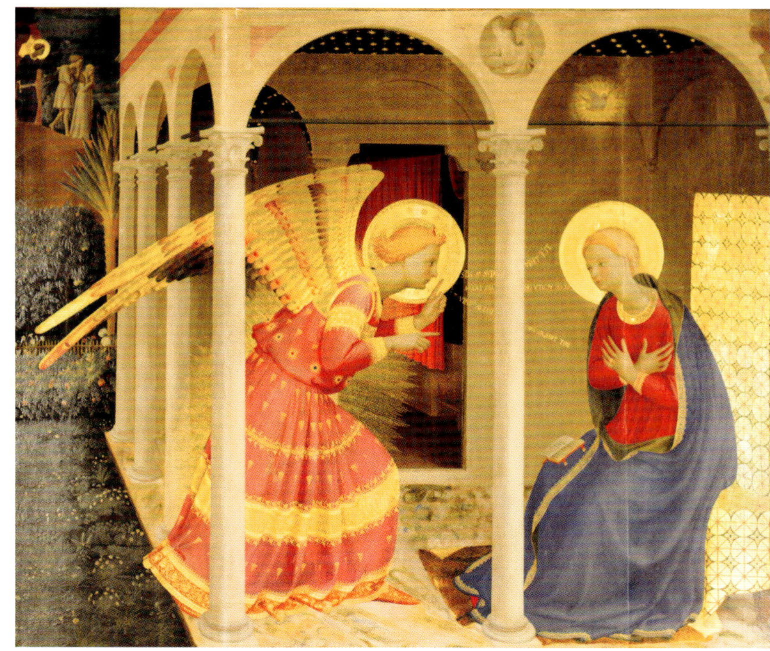

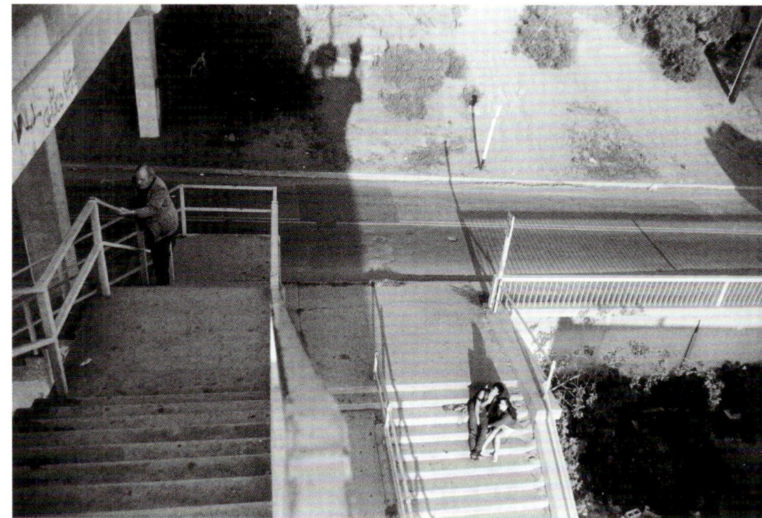

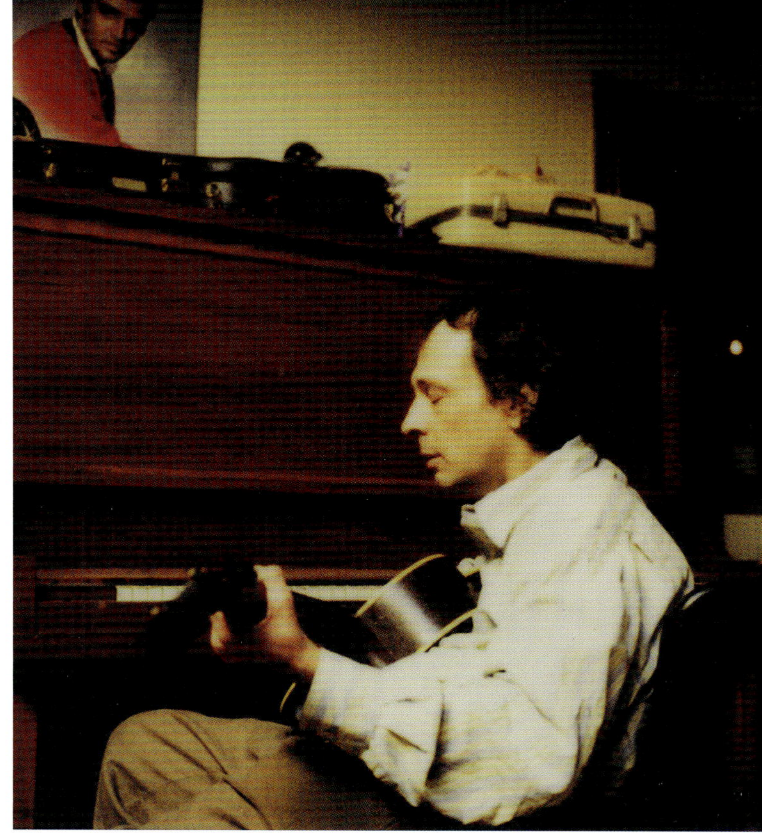

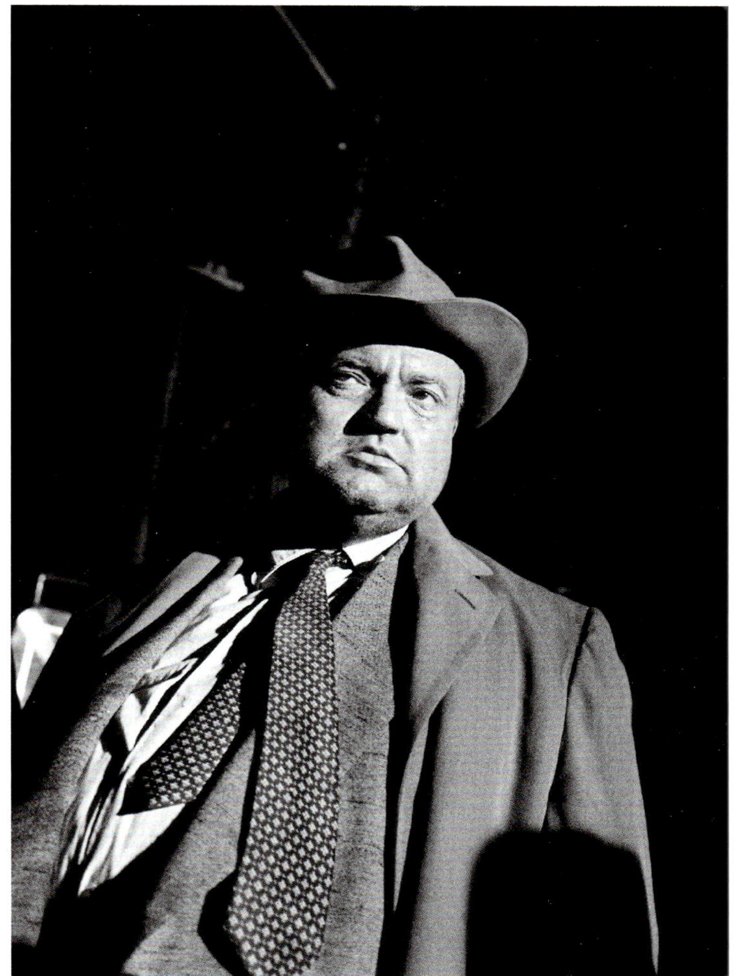

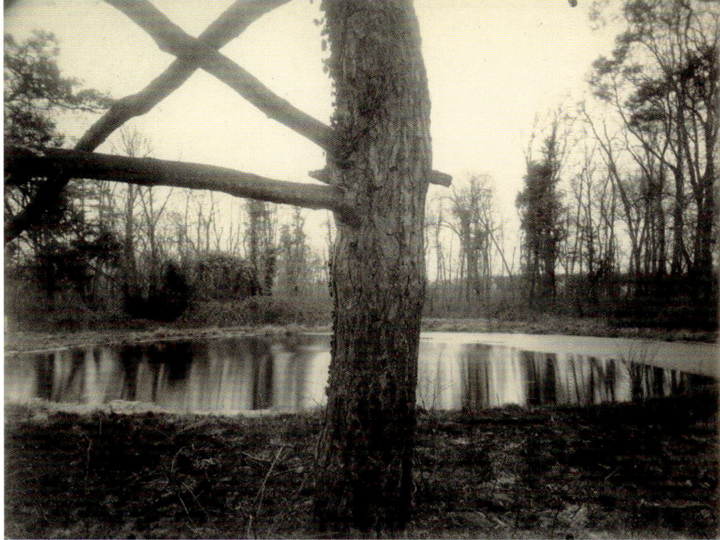

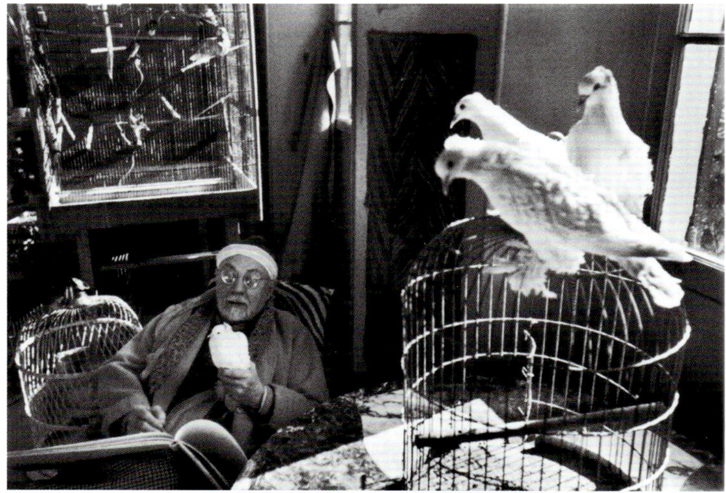

Art | Basel
Hong Kong

Participating Galleries

#

10 Chancery Lane
303 Gallery
47 Canal

A

Miguel Abreu
Acquavella
Aike
Alisan
Anomaly
Antenna Space
Applicat-Prazan
Arario
Alfonso Artiaco
Artinformal
Aye

B

Balice Hertling
Beijing Art Now
Beijing Commune
Bergamin & Gomide
Bernier/Eliades
Blindspot
Blum & Poe
Boers-Li
Tanya Bonakdar
Isabella Bortolozzi
Ben Brown
Gavin Brown
Buchholz

C

Gisela Capitain
Cardi
Carlos/Ishikawa
Chambers
Chemould Prescott Road
Yumiko Chiba
Chi-Wen
Sadie Coles HQ
Contemporary Fine Arts
Continua
Paula Cooper
Pilar Corrias
Alan Cristea
Chantal Crousel

D

Thomas Dane
Massimo De Carlo

de Sarthe
Dirimart
du Monde

E

Eigen + Art
Eslite
Gallery Exit
Experimenter
Selma Feriani
Fortes D'Aloia & Gabriel
Fox/Jensen

G

Gagosian
Gajah
gb agency
Gladstone
Gmurzynska
Goodman Gallery
Marian Goodman
Gow Langsford
Bärbel Grässlin
Richard Gray
Greene Naftali
Karsten Greve
Grotto

H

Hakgojae
Hanart TZ
Hauser & Wirth
Herald St
Max Hetzler
Hive
Xavier Hufkens

I

Ingleby
Ink Studio
Taka Ishii

J

Annely Juda

K

Kaikai Kiki
Kalfayan
Karma International
Kasmin
Sean Kelly
Tina Keng

Kerlin
König Galerie
David Kordansky
Tomio Koyama
Kraupa-Tuskany Zeidler
Andrew Kreps
Krinzinger
Kukje
kurimanzutto

L

Pearl Lam
Simon Lee
Leeahn
Lehmann Maupin
Lelong
Lévy Gorvy
Liang
Lin & Lin
Lisson
Long March
Luhring Augustine
Luxembourg & Dayan

M

Maggiore
Magician Space
Mai 36
Edouard Malingue
Matthew Marks
Mazzoleni
Fergus McCaffrey
Greta Meert
Urs Meile
Mendes Wood DM
kamel mennour
Metro Pictures
Meyer Riegger
Francesca Minini
Victoria Miro
Mitchell-Innes & Nash
Mizuma
Stuart Shave/Modern Art
The Modern Institute
mother's tankstation

N

nächst St. Stephan
    Rosemarie
    Schwarzwälder
Nadi
Nagel Draxler

Richard Nagy
Nanzuka
Taro Nasu
neugerriemschneider
nichido
Anna Ning
Franco Noero

O

Nathalie Obadia
OMR
One and J.
Lorcan O'Neill
Ora-Ora
Ota
Roslyn Oxley9

P

P.P.O.W
Pace
Pace Prints
Paragon
Peres Projects
Perrotin
Petzel
Pi Artworks
PKM
Plan B

R

Almine Rech
Regen Projects
Nara Roesler
ROH Projects
Tyler Rollins
Thaddaeus Ropac
Rossi & Rossi
Lia Rumma

S

SCAI The Bathhouse
Esther Schipper
Rüdiger Schöttle
ShanghART
ShugoArts
Side 2
Sies + Höke
Silverlens
Skarstedt
Soka
Sprüth Magers
Starkwhite

STPI
Sullivan+Strumpf

T

Take Ninagawa
Tang
This Is No Fantasy + dianne
    tanzer
Templon
The Third Line
TKG⁺
Tokyo Gallery + BTAP
Tornabuoni
Two Palms

V

Vadehra
Van de Weghe
Vitamin

W

Waddington Custot
Wentrup
Michael Werner
White Cube
White Space Beijing
Barbara Wien
Jocelyn Wolff

Y

Yavuz

Z

Zeno X
Zilberman
David Zwirner

Insights

A Thousand Plateaus
Asia Art Center
Bank
Baton
Beyond
Dastan's Basement
Don
Empty Gallery
Espace
Fost
Hunsand Space
Yoshiaki Inoue
Johyun
Richard Koh

Mind Set
Pifo
Star
Yuka Tsuruno
Watanuki / Toki-no-
    Wasuremono
Wooson
Yamaki

Discoveries

1335Mabini
A⁺ Contemporary
Sabrina Amrani
Christian Andersen
Capsule
Château Shatto
Commonwealth and
    Council
Crèvecoeur
Ghebaly
High Art
Hopkinson Mossman
hunt kastner
Jhaveri
JTT
Maho Kubota
Emanuel Layr
Michael Lett
MadeIn
mor charpentier
Nova Contemporary
Project Native Informant
Société
Tabula Rasa
Tarq
Vanguard

March 29–31, 2019

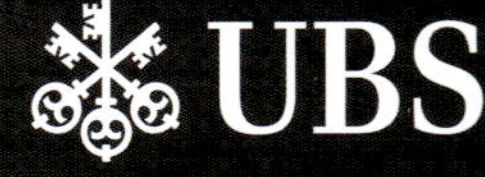

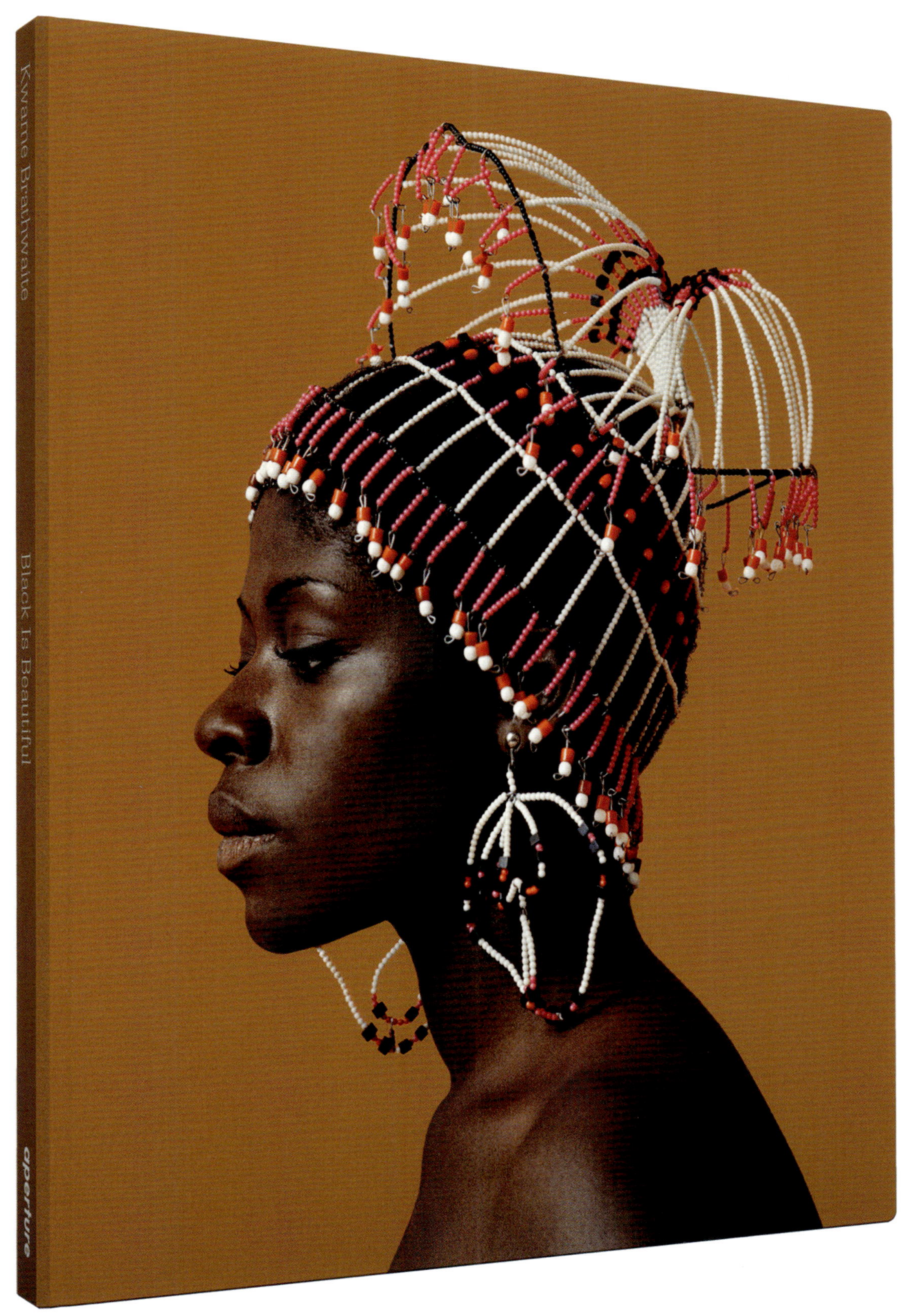

Kwame Brathwaite: Black Is Beautiful

Available April 2019

aperture

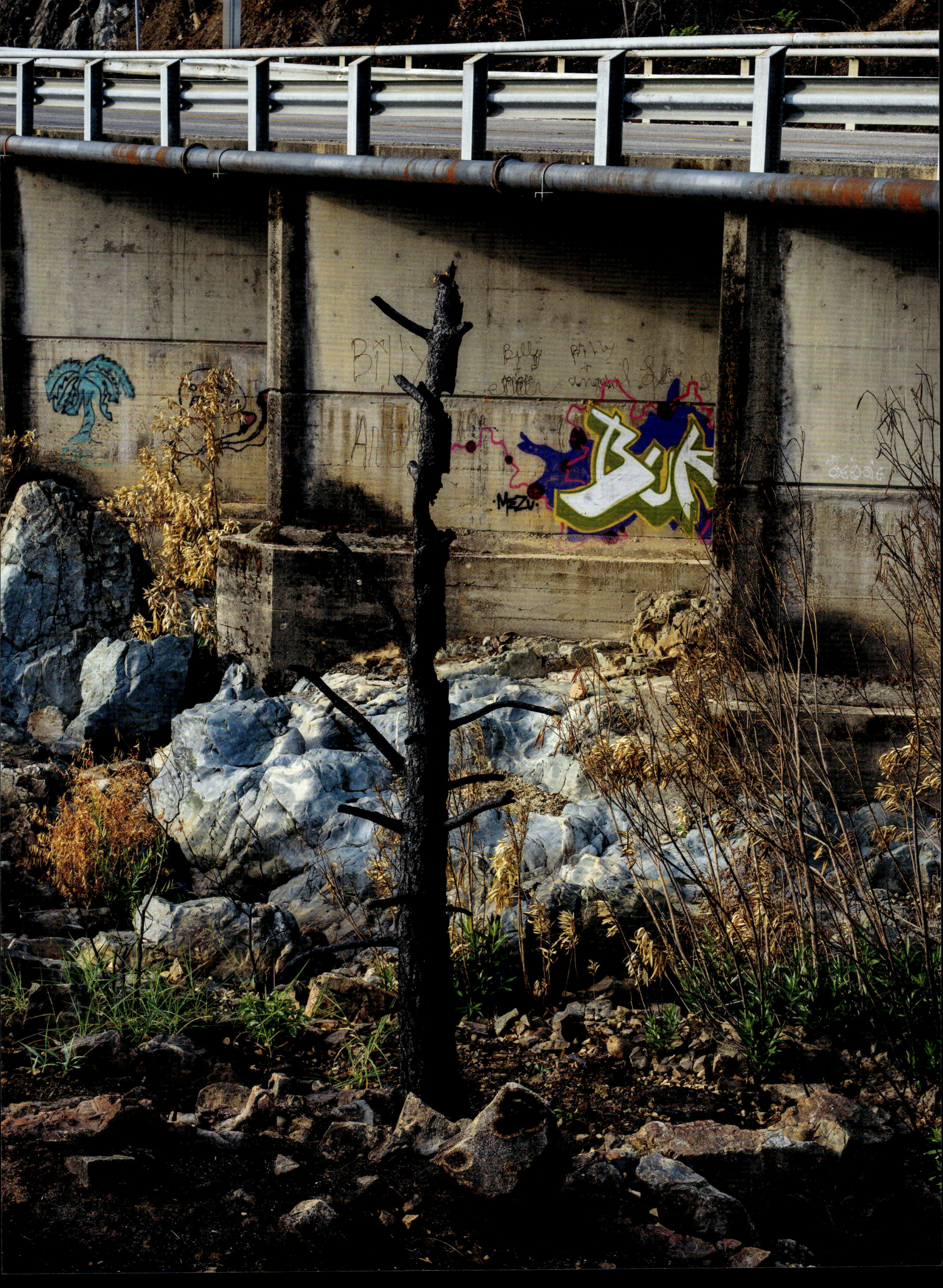

Billy
Billy   Billy
angel
MEZU

Earth

When *Aperture* commissioned Carolyn Drake this past September to photograph the recent wildfires in Northern California, it seemed as if a severe season of destruction had finally concluded. Months passed. As we prepared to go to press with this issue in November 2018, a fresh wave of apocalyptic fires surged across parts of the state, killing dozens, displacing thousands. Hundreds more remained missing. Photographs tend to fall short when faced with such devastation. But Drake's austere images of ravaged landscapes point to the role of people in creating the conditions for this new, terrifying normal. William Finnegan, who was raised in the state, reminds us that we are in uncharted territory. "It is the Anthropocene," he says, referring to our geological age defined by human activity. "We must look to our own agency."

Many of the photographers and writers in this issue do just that. David Benjamin Sherry traveled the United States to document the national monuments reduced in size by the current administration, opening once-protected lands to use by industry. Of Sherry's arresting, analog color that signals an overheated future, environmentalist Bill McKibben observes that we "glimpse that something has gone wrong and is now going wronger in these places." Gideon Mendel salvages found photographs from flood zones around the world, conveying the threat of rising sea levels and extreme weather. Eva Díaz, in her look at art and ecofeminism, discusses three contemporary artists who address climate change and challenge the gendered idea of our so-called mastery over Earth. Wanda Nanibush describes the power of photographs for Indigenous artists, which provide "forms of

visual sovereignty and assert a continued presence on the land, despite centuries of theft and removal." She, along with scholar T. J. Demos, underscores the importance of sociopolitical context when considering how we visualize the environment and questions of ecology.

While planetary life may be in peril, it remains a source of visual fascination and discovery. From Anna Atkins's nineteenth-century cyanotypes of botanical specimens to Karl Blossfeldt's twentieth-century sculptural plant anatomies, photographers have long found inspiration in natural forms and structures. Following in this tradition are Jochen Lempert, a scientist turned artist who photographs myriad plant life, and Bruno V. Roels, who collages landscape images of palm trees with his own markings and cryptic messages.

The organic world also provides sites for mythology and storytelling. Vasantha Yogananthan's hand-painted images retell an ancient Indian epic staged in a mythical jungle that is also an actual topography, one now impacted by contemporary concerns of industry and migration, while Arguiñe Escandón and Yann Gross travel to Peru in search of connections to nature in the Amazon. For Lieko Shiga, land is freighted with memory. *Human Spring*, her new, profoundly original series featured in these pages, grapples with the continued psychological and ecological aftermath of Japan's 2011 Tohoku earthquake and tsunami. Four days after the disaster, Shiga returned to her studio in Tohoku, only to find it had vanished, along with her home. "What does it mean," she asked herself, "to no longer be able to live on your own land?" —**The Editors**

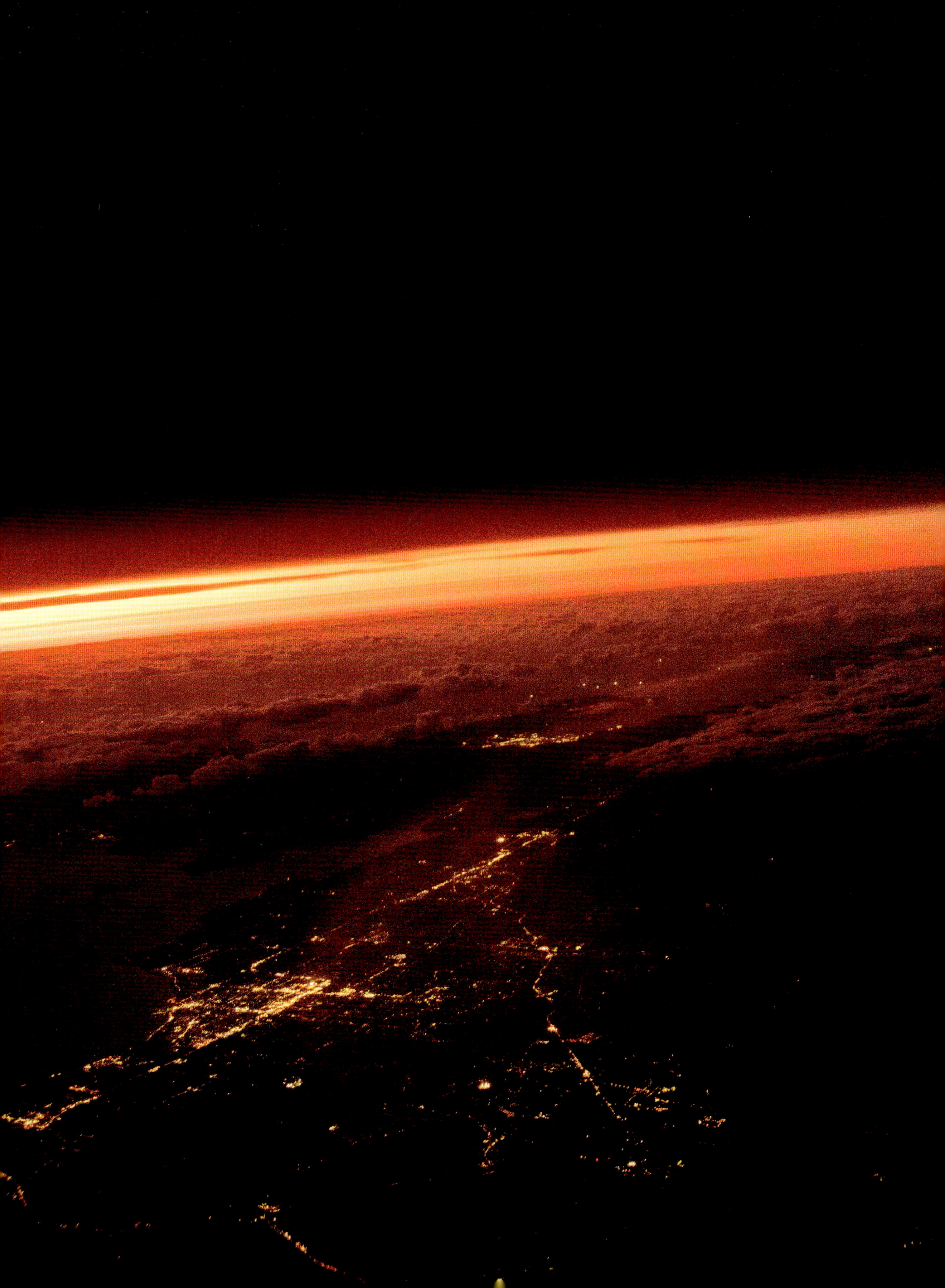

LIEKO SHIGA

Eight years after a devastating tsunami, an artist investigates
Japan's haunted landscapes
Amanda Maddox

HUMAN SPRING

Death, like the tsunami about to usher it ashore, moves toward her. She is standing unaware outside a grocery store in the city of Natori. She can't see the ten-foot waves cresting in the Pacific Ocean. She doesn't feel the seismic plates shifting, scraping, subducting, or lifting the sea floor approximately eighty miles away. All she knows is what they told her: tsunamis do not happen in Kitakama village, where the sandy coastline near her studio runs straight and narrow. Then the Great East Japan Earthquake hits on March 11, 2011, the first blow of the "triple disaster" that will claim more than 15,500 lives. But photographer Lieko Shiga, alone in a parking lot, twenty miles from the village, doesn't realize this yet.

When the aftershocks subside, Shiga runs inside the store. Everyone is safe. She returns to her car and drives toward her studio—a prefab trailer nestled among the magical grove of pine trees adjacent to Kitakama beach. Her boyfriend calls, unscathed but upset. "Please don't go to your village," he says, having heard a tsunami warning issued over the radio. "It's coming." But she keeps driving until, suddenly, she sees it. The sea stretches out like a long, dark coffin. A massive, brown wave filled with debris barrels toward her white Mazda. She reverses quickly and races in the opposite direction, keeping one eye on the rearview mirror as the waves close in behind her. Along the road that leads inland, she passes a mother with her son walking calmly and obliviously. Shiga doesn't know them, but she urges them into the car, where they wait out the snowy night together parked in front of Natori City Hall.

I first heard Shiga recount parts of this story in summer 2014. We were in Tokyo, and she was pregnant with her son, Shiita. As she recalled the nearness of the waves, her protruding belly came to signify her survival. She embodied the notion that proximity to death conversely breeds proximity to life. Similarly, hearing the tsunami described as more devastating than war, she told me, made her identify that event as the first "super real" experience of her life. Born in 1980 and raised in a suburb of Okazaki in Aichi Prefecture, where the neighboring Toyota plant and assorted manufacturing facilities employ the majority of the local workforce, she always perceived her immediate environment as an empty stage, with props and puppets acting out scenes before her eyes. But in the aftermath of the disaster, she "can finally feel what's happening in the world." This sensation comes to bear in her work. "I can imagine quite clearly now. I can trust my imagination. And it's not only about dreaming; I'm connecting."

In 2008, Shiga moved to Kitakama in Miyagi Prefecture, motivated in large part by the promise of connection. A residency in nearby Sendai two years earlier introduced her to this place—its peculiarities, its agrarian roots, its strong and intoxicating natural landscape. She was immediately entranced. "I wanted to jump inside the photographs I was making here," she said. After seven years in London, where she studied photography at Chelsea College of Art and Design until 2004, followed by two years in Berlin and residencies in Singapore and Brisbane, Shiga boomeranged back to Japan for good. Kitakama had emerged as "the place I'd been looking for all along."

How could this village, situated in the "backwater" region often regarded as "old Japan" because of its slow pace and rural lifestyle, be part of the same country she called home? Like many people, her knowledge of northeastern Japan—six prefectures collectively known today as Tohoku (meaning "northeast") but formerly called Michinoku, meaning "end of the road" or "back roads"—derived almost exclusively from Kunio Yanagita's *The Legends of Tono* (1910).

But that folkloric account represents one academic's journey across Tohoku, she is quick to point out. He purportedly spent little time in Miyagi Prefecture before returning home to Tokyo. Shiga's initial impression of the area as silent, minimal, and imbued with a punk spirit also diverged from Yanagita's account. Intrigued and determined to discover this unusual place for herself, she understood that as an outsider she needed to establish her studio there.

To simultaneously announce her arrival, in January 2008, and ingratiate herself among villagers, Shiga passed out flyers that read: "Pleased to meet you. My name is Lieko Shiga. I have rented a space in the pines next to the Kitakama Pool … I am a photographer, and I usually travel around to various places and take photos. This will be my center of activities for the new year…. I will do my best to serve you." She became the local community photographer. In this role, she documented everything from baseball games to town hall meetings, as well as festivals and ceremonies on the verge of dying with members of the aging population. She made portraits that villagers could give to family members or use for their own funeral services. As word spread, people began to visit her studio out of curiosity. Soon she began recording their oral histories, collecting information about them, the local environment, the economy, and the history of the village. A complex portrait of Kitakama consequently emerged, leading her to believe: "Japan is in Tohoku."

Four days after the tsunami hit, Shiga returned to her studio for the first time postdisaster. It was gone. She couldn't even locate the plot of land it once occupied because sand stretched across the forest like a massive tatami. The home she rented nearby had also washed away. "What does it mean to no longer be able to live on your own land?" she asked herself. By that point, fifty-three of Kitakama's 370 residents had been confirmed dead; another seven were declared missing. Shiga had photographed all of them. How do you pick up the pieces of projects interrupted by disaster and death? Can you continue? Do you start over?

She arrived at an answer quickly, which she articulated in an email to worried friends and family on April 5, 2011: "What I feel compelled to confirm with my entire being is that what I started … is not over at this moment. If anything I have done in Kitakama up to now was rendered meaningless by the disaster, it was just the things that could be washed away." But she didn't miss her camera. In fact, she struggled with feelings of disappointment with the medium of photography—years earlier, it had allowed her to discover her existence, but now it seemed empty. And yet, charged with a sense of responsibility as the community photographer, she borrowed a camera to record her neighborhood. The choice to remain busy and connected, to serve this place as a documentarian, defined this period in her life. It informed her decision to voluntarily live in temporary shelters for three years rather than move in with her boyfriend and his family, whose home was not damaged. It justified her participation in the cleaning of found photographs at the community center. It gave her a reason to live. It alleviated pain. It allowed her to forget what happened. It allowed her to remember.

Shiga has since realized two major projects; both embrace the past and explore notions of memory. In 2012, she released *Rasen Kaigan*, a body of work rooted in Kitakama that concerns the landscape—its physicality, its mythos, its history, and human intervention into it—and constitutes a collaboration with local residents. Last year, she revisited and updated a project, from 2008, called *Blind Date*, which features photographs of couples on motorbikes in Bangkok. The tsunami claimed most photographs made prior to March 2011 that Shiga envisioned utilizing in these series, but some files stored elsewhere remained accessible. Trusting that the pictures had been left behind for a reason, she incorporated them into new work. In the case of *Rasen Kaigan*, she also felt it necessary to reconceptualize the project on account of the tsunami. But the disaster itself would not become her subject; it was one moment that violently punctuated her extended engagement with the community and the land of Kitakama. What she needed to express was how the tsunami had entered her body.

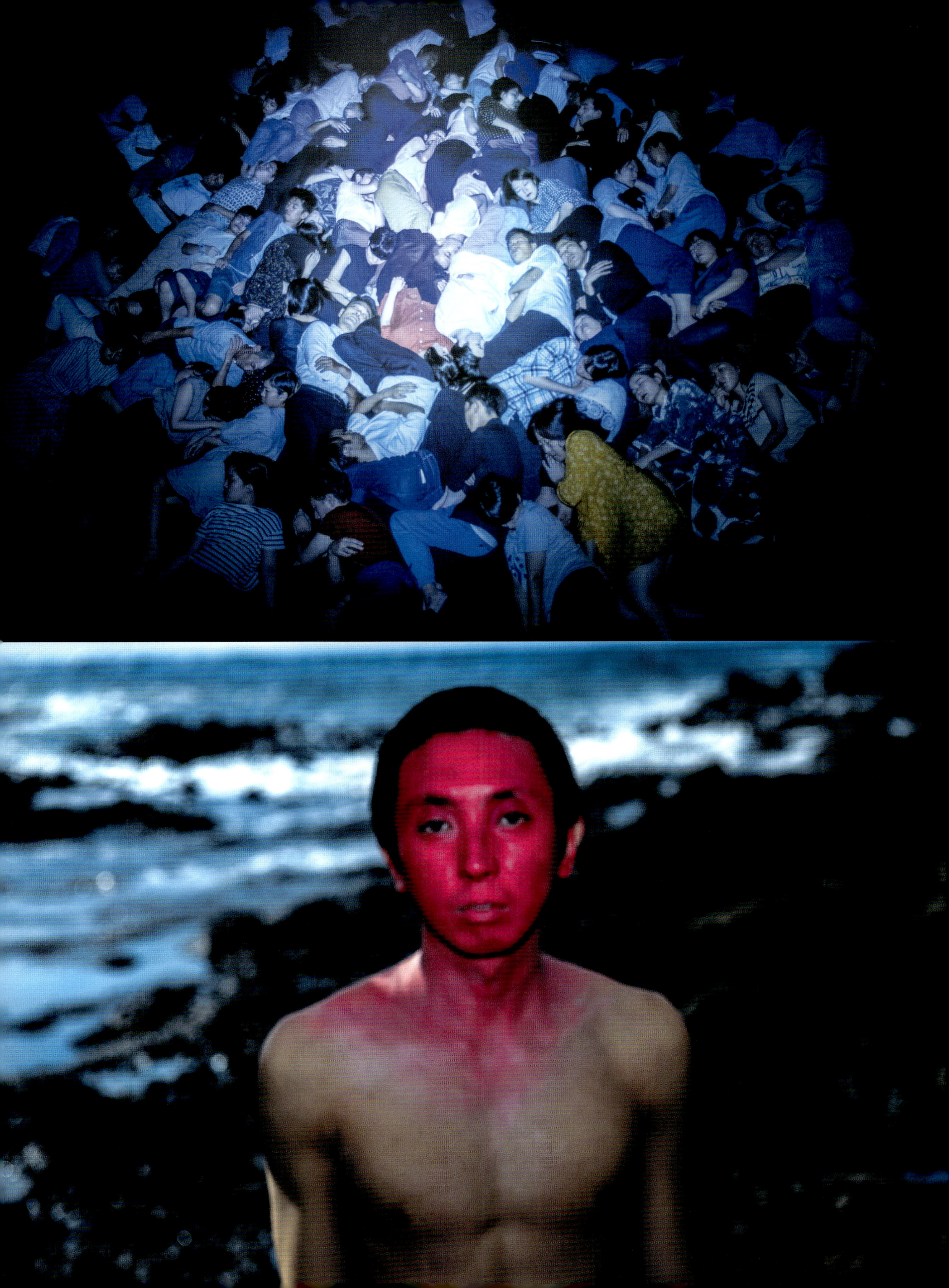

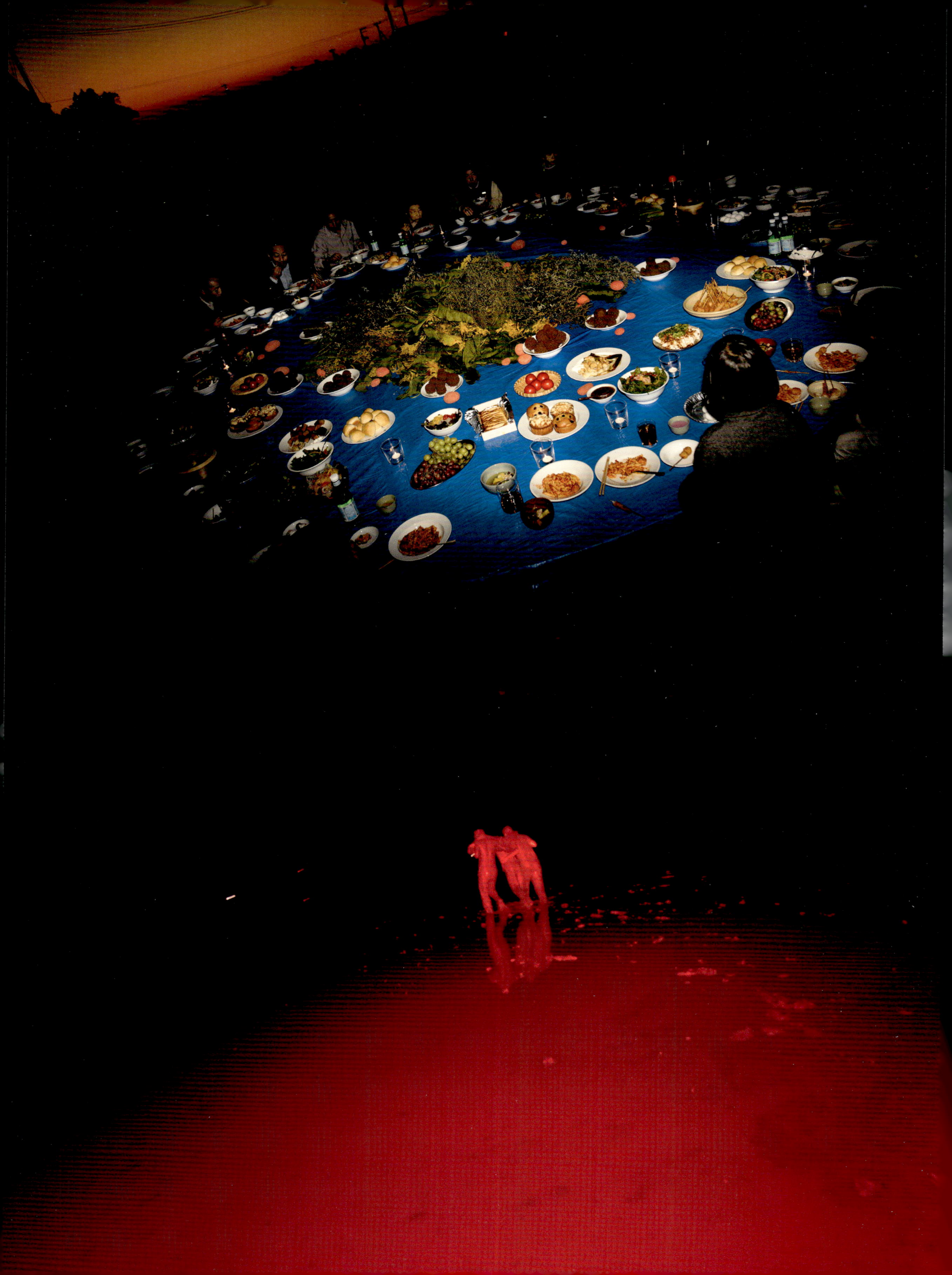

Shiga has alternately referred to herself as a camera, a vessel, and a medium. In all of these descriptions, photography takes a physical form. This perspective stems from her training in dance, which ended abruptly at the age of eighteen. Aware and deeply self-conscious of how her body was rapidly changing, Shiga abandoned dance for photography in high school, a decision she later questioned upon encountering the experimental, expressive work of Pina Bausch. However, she ultimately found she could tap her interest in performance and physicality as a photographer. This often involves pushing her body into the frame, whether literally or figuratively. In *Rasen Kaigan*, one self-portrait exemplifies this idea: *map showing the relationship between photography and myself* (2012) depicts her carving concentric circles into the sandy beach with the root of a pine tree. But arguably the most personal incursion in the series occurs in the photographs of white stones, something Shiga likens to mirrors that reflect memories of faces, animals, or trees found in Kitakama. Upon viewing the exhibition of *Rasen Kaigan* at Sendai Mediatheque, Miyagi Prefecture, in late 2012—where prints were displayed as a dizzying gyre in a darkened room—visitors responded to the sight of these enigmatic stones as she had. Many brought their own associations. Everyone saw Kitakama. Critic Minoru Shimizu, who took the minority position when he dismissed her photography as the stuff of "B-grade horror," declared the installation "cleverly constructed as a stage device in which the acts of recalling and mourning alternate." Former Sendai Mediatheque curator Chinatsu Shimizu said the exhibition resembled a funeral, with prints angled upright like tombstones in a cemetery, and the visitor experience akin to bon odori, a traditional style of dancing during the Obon festival in which people dance in circles to mourn the dead.

※

In *Spring Snow* (1969), Yukio Mishima writes, "There is an abundance of death in our lives. We never lack reminders—funerals … memories of the dead, deaths of friends, and then the anticipation of our own death." The proliferation of death, alluded to in *Rasen Kaigan*, spurred Shiga's new body of work, *Human Spring* (2018–19), the third part of a trilogy that includes *Rasen Kaigan* and *Blind Date*. In 2012, while living at an evacuation center near Kitakama, Shiga found herself surrounded by mortality. Two neighbors, residing in the apartments that flanked hers, died by suicide during her stay. They were the only individuals there to fall victim to what the Reconstruction Agency of the Government of Japan officially termed "disaster-related deaths." Classified as indirect damage caused by either physical or psychological stress, such casualties triggered by the earthquake and tsunami emerged collectively as a crisis. Many of those affected, including one of Shiga's neighbors, were farmers who resorted to suicide because their oversalinized crops could not be sold.

But the farmer whose story anchors *Human Spring* is S-chan (whose name has been changed to protect his family's privacy). A native of Kitakama, S-chan grew Chinese cabbage and melons on land owned by his family for many generations. In the late 1980s, a government-supported initiative to stimulate further agricultural production in Tohoku—an area historically cursed by famines and harvest failures—allowed him to install a greenhouse. Eager to see how his crops would perform in this new environment, S-chan visited his farm obsessively until one morning, when he discovered that everything had died overnight. He eventually recovered and revived his business, but the seed of depression had been planted inside him. It germinated for decades.

Shiga met S-chan years later, in winter 2008. He often stopped by her studio during his morning walks to the beach. They exchanged pleasantries and became friends. But within a few months, Shiga

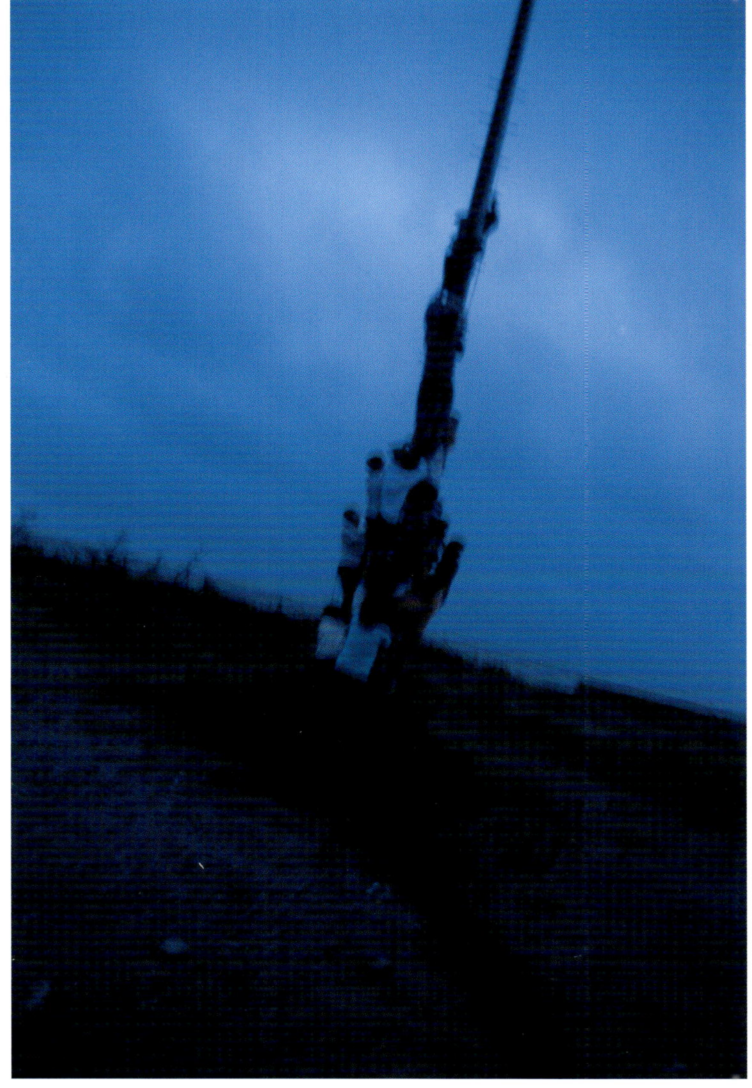

All photographs from
the series *Human Spring*,
2018-19
Courtesy the artist

Shiga has alternately referred to herself as a camera, a vessel, and a medium.

recognized something odd about his character, something that everyone noticed but no one acknowledged out loud. In spring, about forty-eight hours before the cherry blossoms bloomed, S-chan transformed. His cheeks turned red. He never slept. He drove his tractor around town in the dark. He talked incessantly. According to Shiga, S-chan's particular strain of mania always coincided with the change in seasons; his "super high" marked the onset of spring, which arrives so suddenly in Tohoku that "the body must respond before the heart," Shiga said. "He was the spring itself."

Immediately after the tsunami, Shiga and S-chan spent three months in the same temporary shelter staged inside a local middle-school gymnasium—a location referenced in *Human Spring*. His strange behavior continued there. He walked around clutching a Jizo statue, like a child with a stuffed animal. He visited the beach often to confer with the ghosts because, he claimed, they were lonely. Every morning, he flung open all the curtains at sunrise to announce the arrival of a new day. Though this annoyed practically everyone at the shelter, they were so charmed by S-chan's idiosyncrasies and positive energy that "springtime mania infected them, too," said Shiga.

In 2013, soon after the suicides in the evacuation center, S-chan succumbed to cancer. Shiga couldn't help but draw a parallel: one death prompted by disease, two deaths instigated by societal and financial pressures, all of the men victims of the Heisei depression—a reference to the Imperial era during which they died—and all connected through her. As time passed, she kept thinking about their final moments and thoughts, the state of their corpses, and their potential reincarnation. After these tandem deaths, Shiga confronted the fragility of her own existence when diagnosed with hyperthyroidism. She lost control of certain faculties, she sensed S-chan's manic spirit pulsate through her every time her metabolism accelerated, and she started to channel this energy into *Human Spring*.

For Shiga, the process of generating her latest series began by positing simple but expansive questions: What is the history of human expression? What is inside the human body, and how is it connected to the earth and to society writ large? Such queries demanded extensive research and fieldwork, which, along with the active participation of her subjects, are mainstays of Shiga's practice. She draws inspiration from everywhere: friends, dreams, films, dance, news events, nature, literature, mythology, personal experience. The concept of spring—as the season when cherry blossoms both bloom and die, as a period that has defined such historic uprisings as the Prague Spring and the Arab Spring, as a physical movement—serves as an organizing principle. Sometimes these varied influences and ideas yield highly controlled surrealistic work, from sculptures made in her studio that evoke Mike Kelley's *Kandors* series (1999–2011) to a choreographed scene in the landscape of a suspended forest, yet her process never involves digital manipulation.

But in alignment with Shiga's own delicate physical state, *Human Spring* reflects and arises from discomfort and intentional disorder. An image of the off-kilter horizon, made from an airplane window, encapsulates the imbalanced, dualistic spirit of the project. She also wrestled with personal and national trauma by photographing in Fukushima Prefecture, where she focused on the ghost town of Futaba (population zero), located five miles from the Daiichi Nuclear Power Plant, which released radioactive material into the ocean, soil, and air in 2011. And unlike in previous series, Shiga worked with many anonymous people and sought unfamiliar locations while making these images. These circumstances destabilized certain preconceived ideas of the project and even disrupted the previsualized images she sketched out on paper, forcing a loss of control, much like the megathrust earthquake shifted Earth's axis and threw off its orbit. Only

under such uncomfortable conditions, she claims, could she make *Human Spring*.

Conceived as a series of personal pictures, *Human Spring* originally bore the title *S-chan's Spring* until Shiga realized that his story represented a set of more universal, interconnected social, environmental, and economic fault lines running just below the surface of things in Japan. S-chan's depression felt deeply intertwined with the development of the Heisei era, which commenced in 1989 not long before the economic bubble collapsed in the early 1990s, around the time his greenhouse was installed. Broadly defined by its excess and emptiness, the Heisei period will come to a close on April 30, 2019, when the emperor abdicates his throne. In his 2017 book *Ghosts of the Tsunami: Death and Life in Japan's Disaster Zone*, journalist Richard Lloyd Parry observed that the country has been "adrift, becalmed between a lost prosperity and a future that was too dim and uncertain to grasp" during much of the Heisei era; though the disaster, in 2011, had the potential to jolt "Japan out of the political and economic funk into which it had slithered," this has not occurred. Perhaps the presentation of *Human Spring* at the Tokyo Photographic Art Museum, in spring 2019, scheduled to overlap with the end of the Heisei era, will usher in a new wave of reflection, with images that defy any sense of calmness and complacency associated with the past three decades.

As Tohoku continues to recover and rebuild, the land still registers wounds of contamination, devastation, and depopulation. "Everywhere is the shadow of death," as Rachel Carson wrote of a fictional town sprayed with DDT in *Silent Spring*, her classic 1962 book about the effects of toxins in the environment; this valuation could apply to many villages in northeastern Japan today. Citing Carson's influence on *Human Spring*, Shiga acknowledges, "If I want to capture what's happening when we're alive, I have to think about death." It is omnipresent, it is internal. Ideas about life and death swirl inside her constantly and quickly, such that "her mind is a kind of ocean," as her publisher Tomoki Matsumoto put it. The tsunami that seeped into her body now spills out in her work. And one side effect, Shiga says, is that "I am living with a lot of dead and I may be a ghost."

Amanda Maddox is associate curator, Department of Photographs, at the J. Paul Getty Museum, Los Angeles.

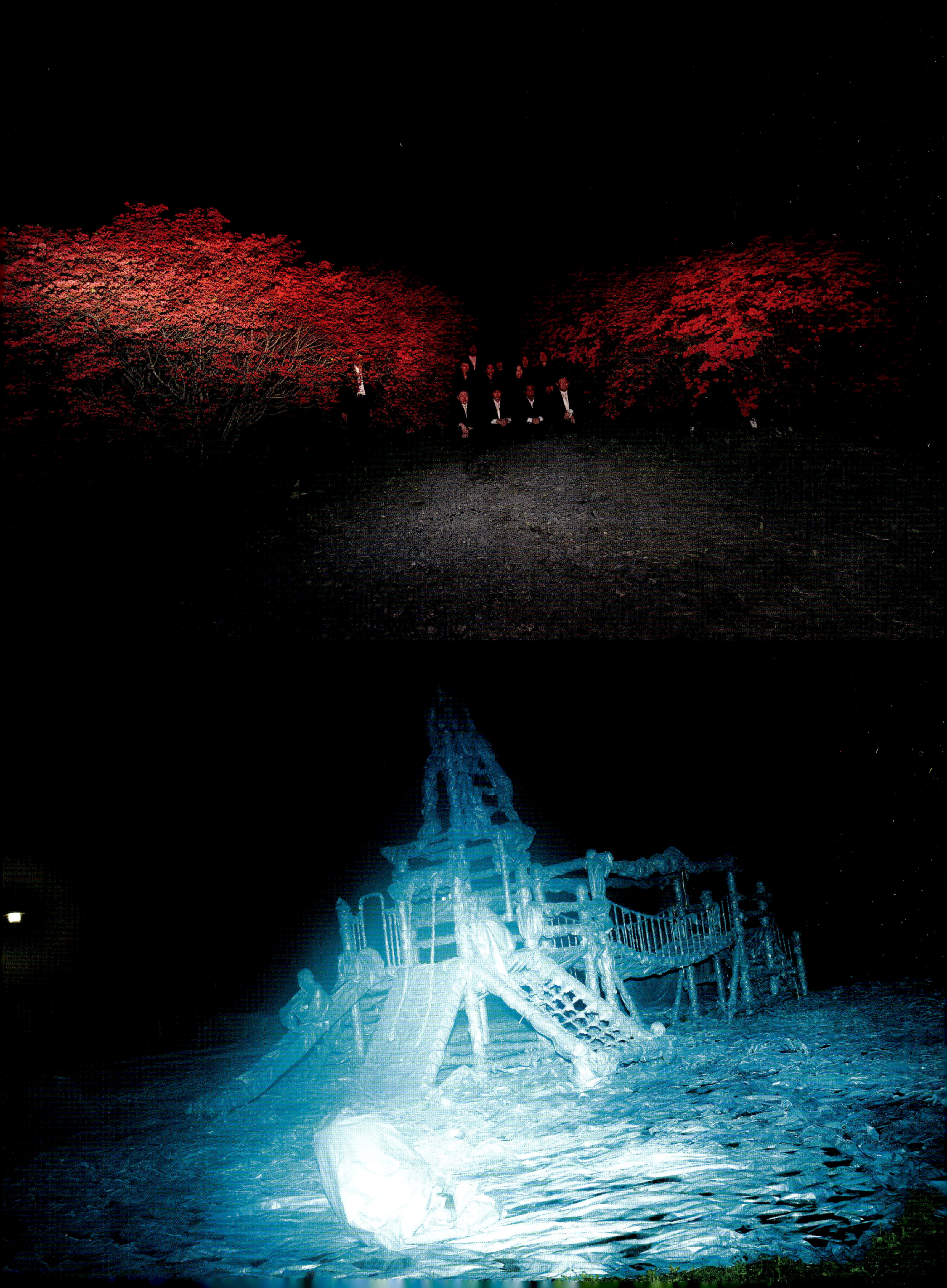

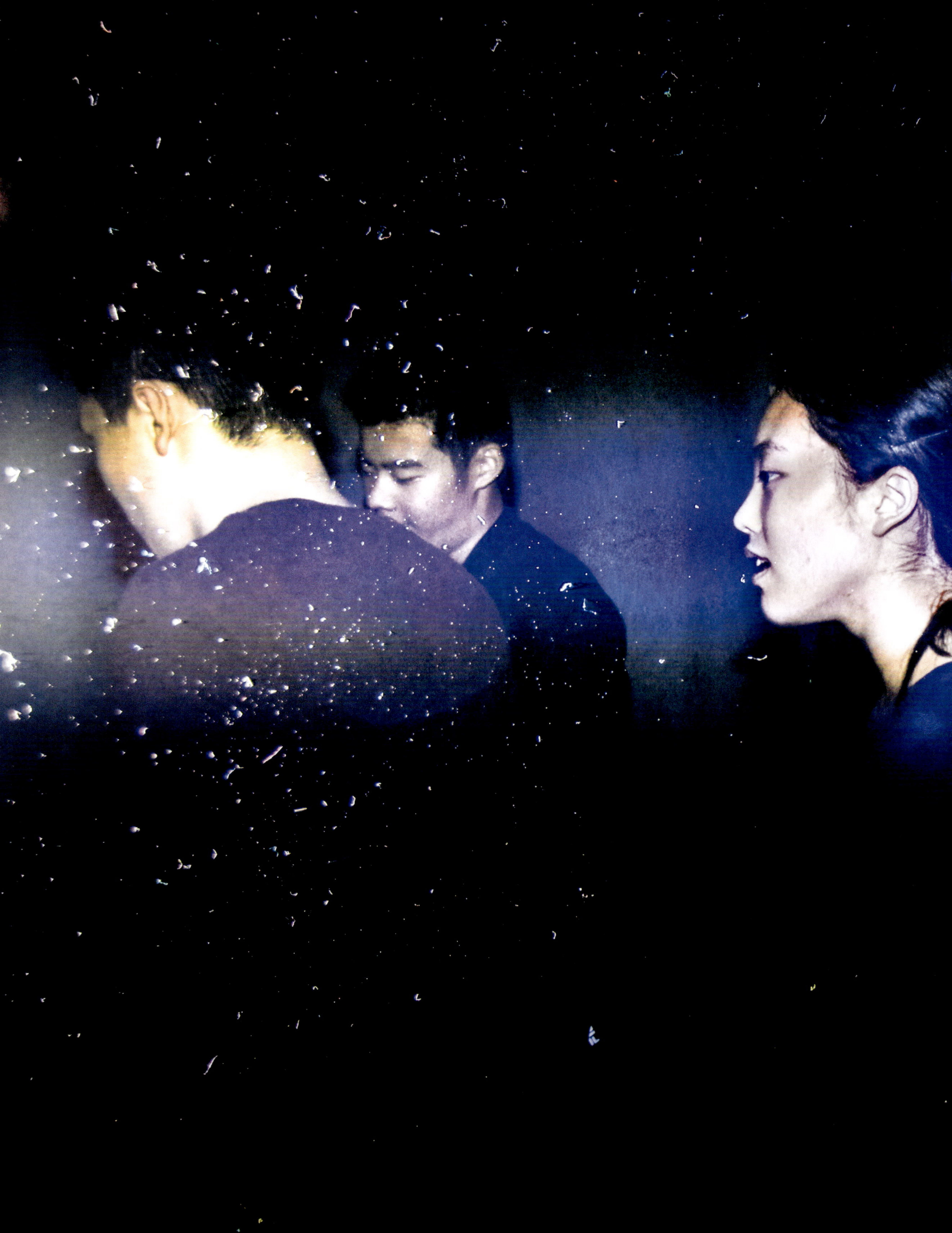

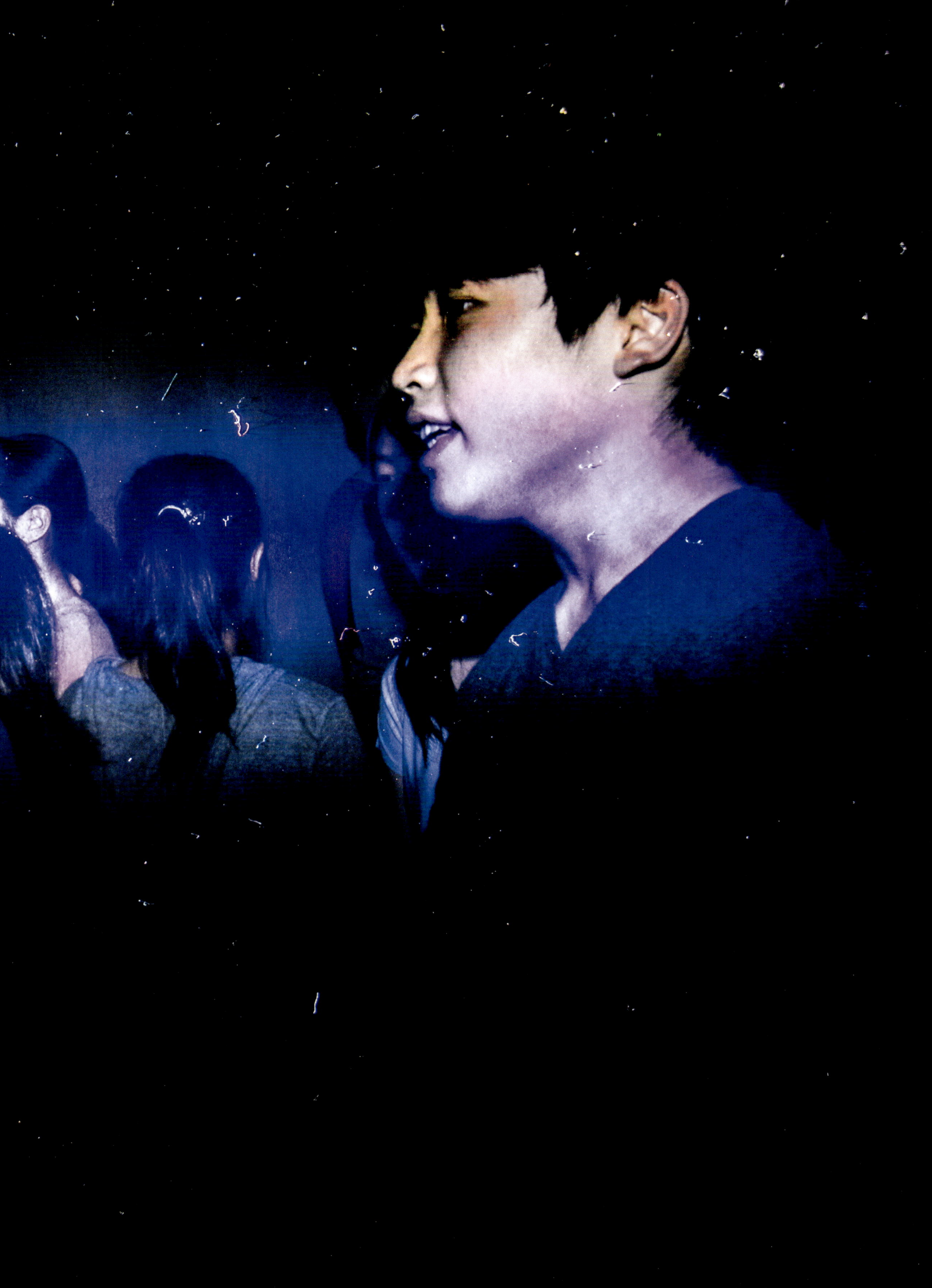

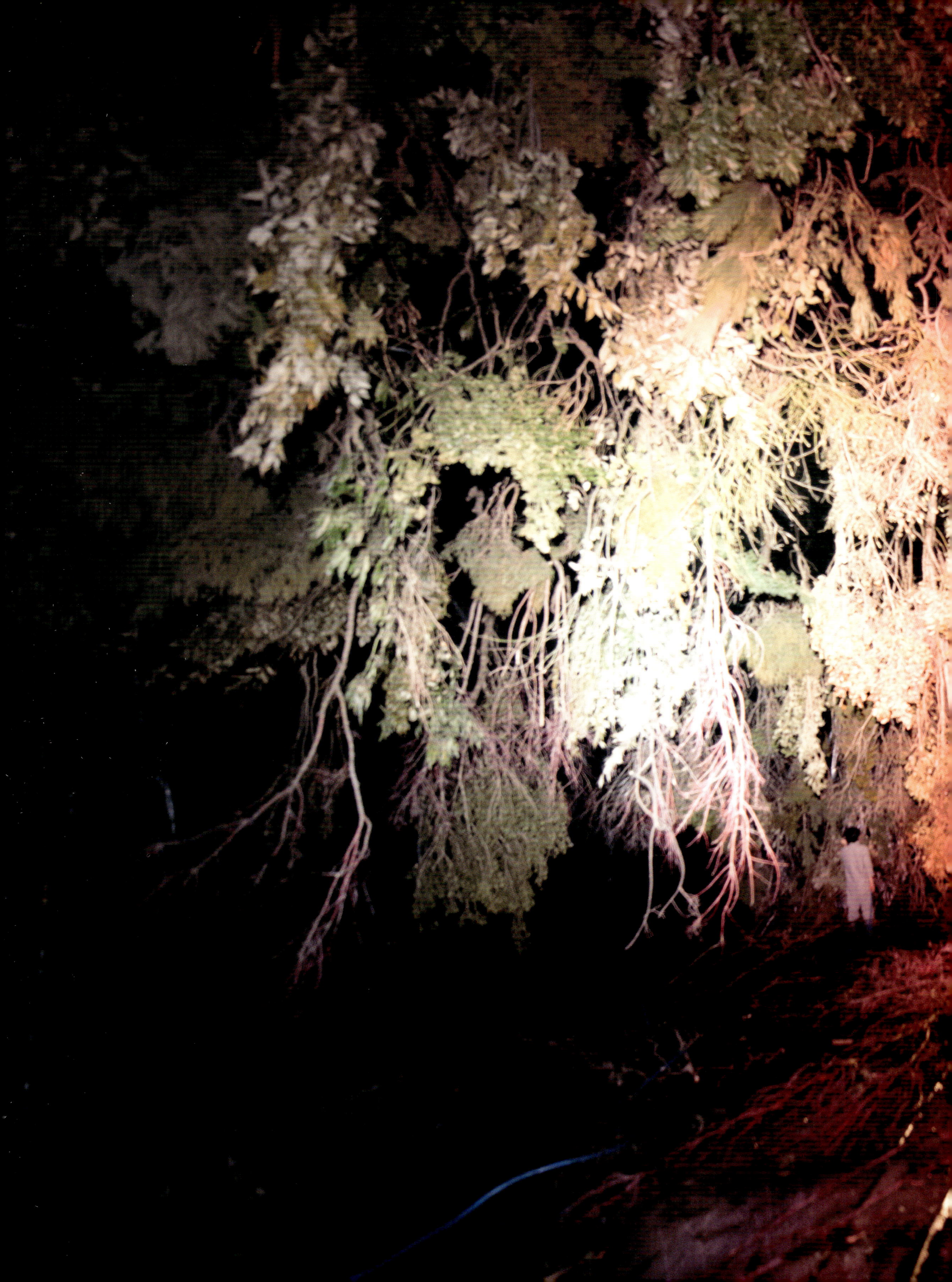

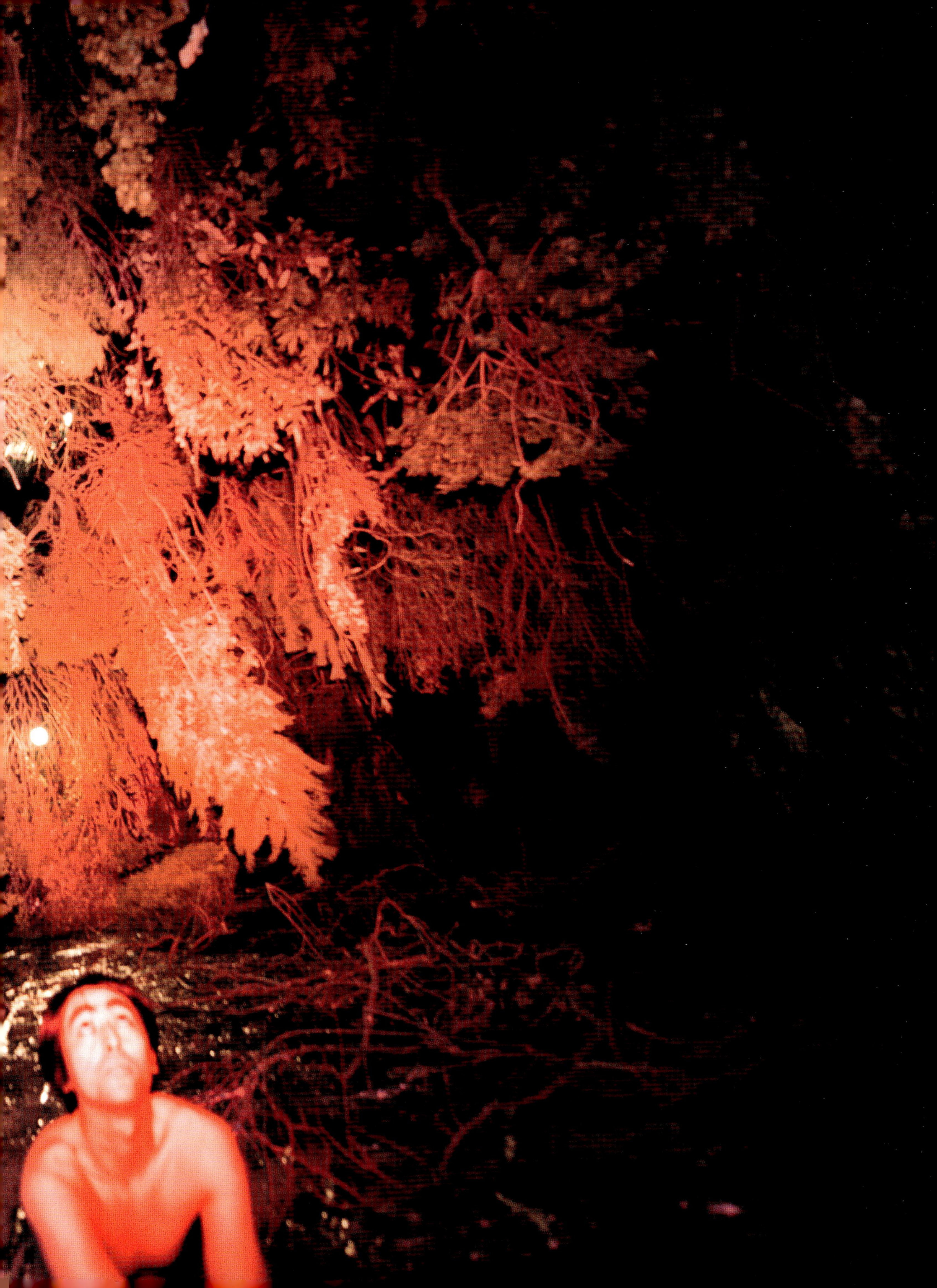

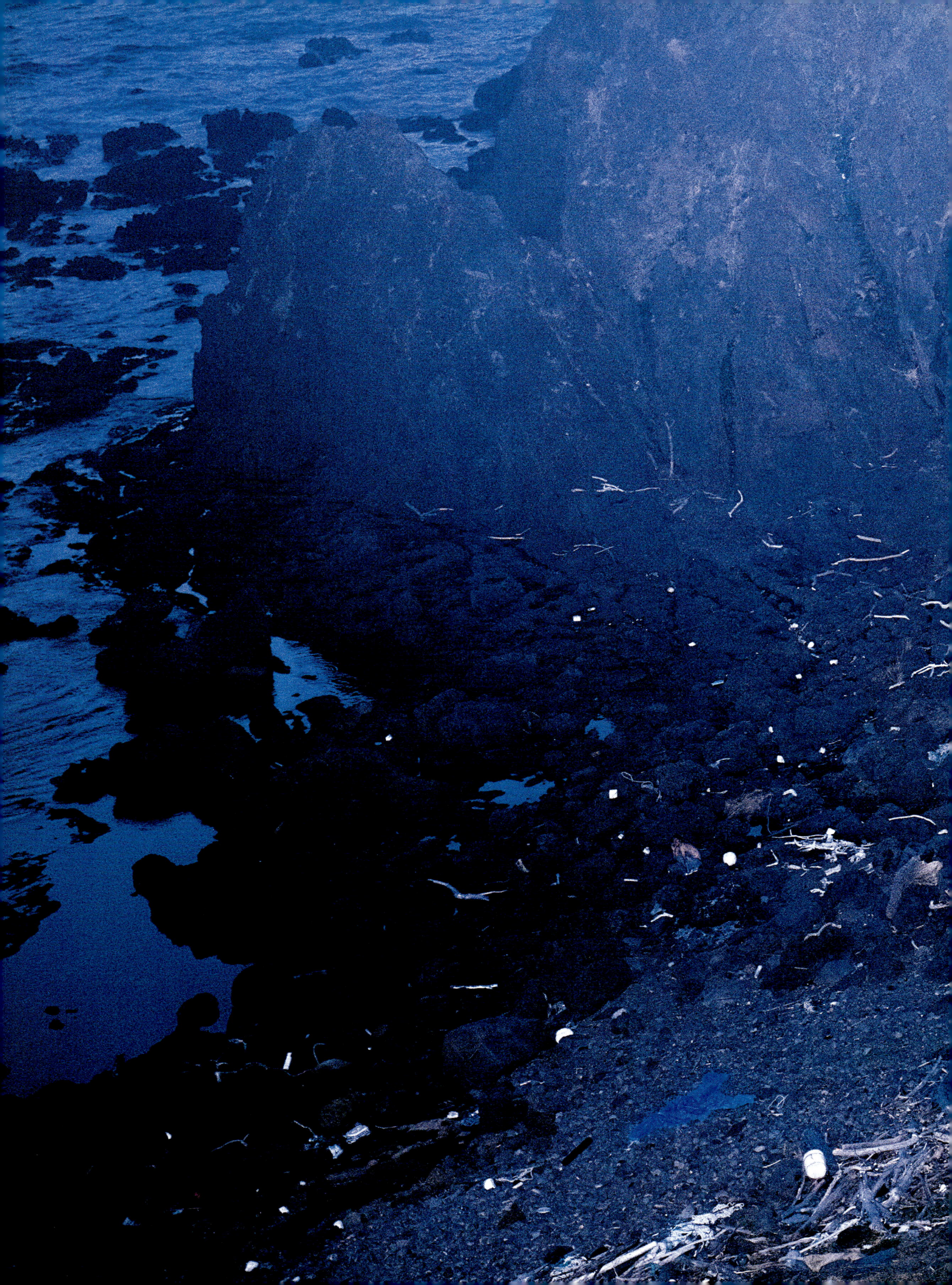

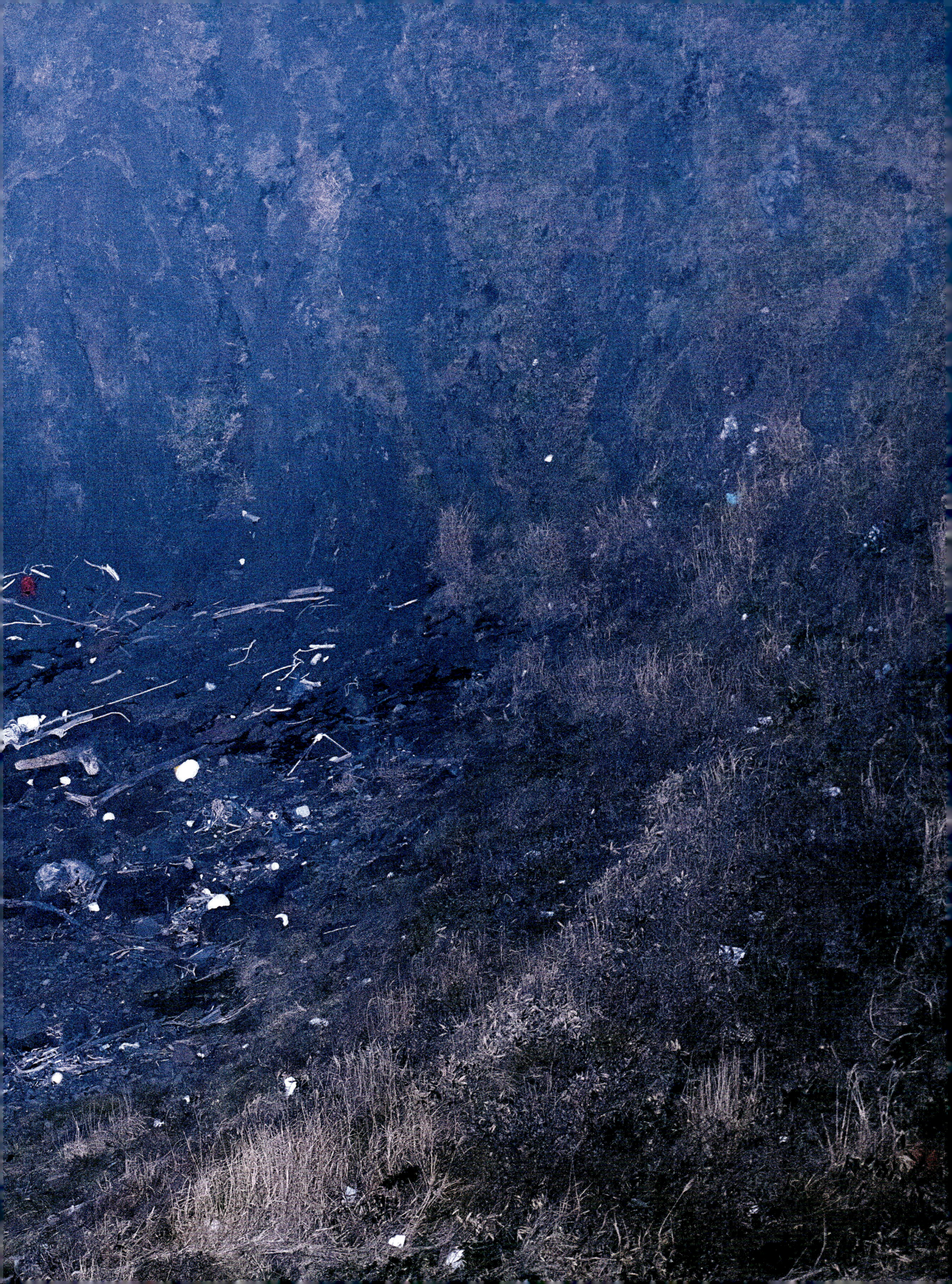

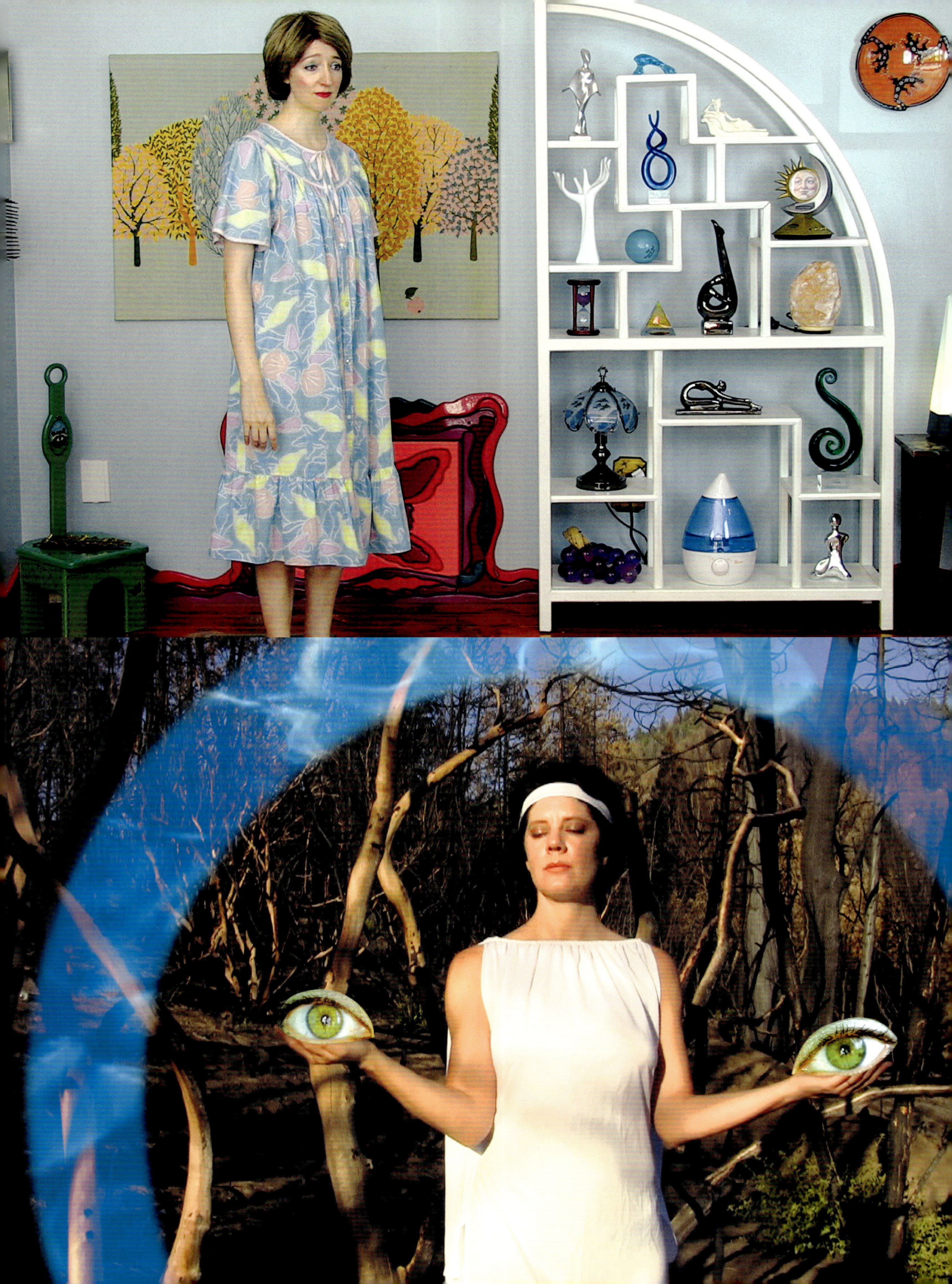

Ecofeminist World Building

Three artists respond to the urgent crisis of climate change
Eva Díaz

Undoubtedly the land, oceans, and atmosphere of Earth have been forever changed by human technologies, in ways that humans cannot roll back. For philosopher Michel Serres, this presents a paradox of ineffective power. As he once noted in a conversation with fellow philosopher Bruno Latour, "We are now, admittedly, the masters of the Earth and of the world, but our very mastery seems to escape our mastery.… Everything happens as though our powers escaped our powers—whose partial projects, sometimes good and often intentional, can backfire or unwittingly cause evil."

The sense of helplessness Serres writes about can easily become a kind of inertia. How to act in the face of the continuing corporate-driven exploitation of Earth, happening at a pace and on a scale that countercollectives of environmentalists, scientists, and the everyday you or me have done little to abate? Artists are in a particularly challenging position with respect to climate change: giving form to these largely invisible forces of capital and their effects can be difficult. And if they do produce work "about" climate change, the almost impossibly complicated matrix of long-term planetary shifts—global warming, rising seas, species extinction, environmental degradation—can make it seem that imaging destruction is yet another elegy for Earth.

In their recent work, three American artists—Shana Moulton, Mary Mattingly, and Connie Samaras—approach human-authored climate change as intertwined with artistic responsibility. They work not merely to provide visual evidence of how climate change manifests itself in often catastrophic ways; these artists act self-reflexively, exploring the role of image making at a moment when humans have perhaps made too many images. Investigating the edges or margins of places, they produce works that fracture and reassemble, more than embody, the effects of human-altered ecologies in what one could term an ecofeminist spirit. In this, they dismantle the gendered notion of "mastery" over Earth to provide alternative visions of the planet, speculating on worlds to come that are often as dystopian as they are optimistic.

For Shana Moulton, this means exploring the dissolving edges of the human and natural. Her film *Whispering Pines 10* (2018) is a thirty-five-minute fantastical journey made in collaboration with musician Nick Hallett. It follows the misadventures of a woman who identifies, to the point of psychic dislocation and death, with tree sitter Julia Butterfly Hill, notorious for camping in the high limbs of a redwood in Northern California for two years to prevent old-growth logging in the region. In the film, Moulton plays the character Cynthia, who appears in many of the artist's works. Cynthia lives alone in the Sierra Nevada mountains in California and is prey to many real and perceived illnesses that she fretfully views as forms of environmental toxicity. She constantly turns to

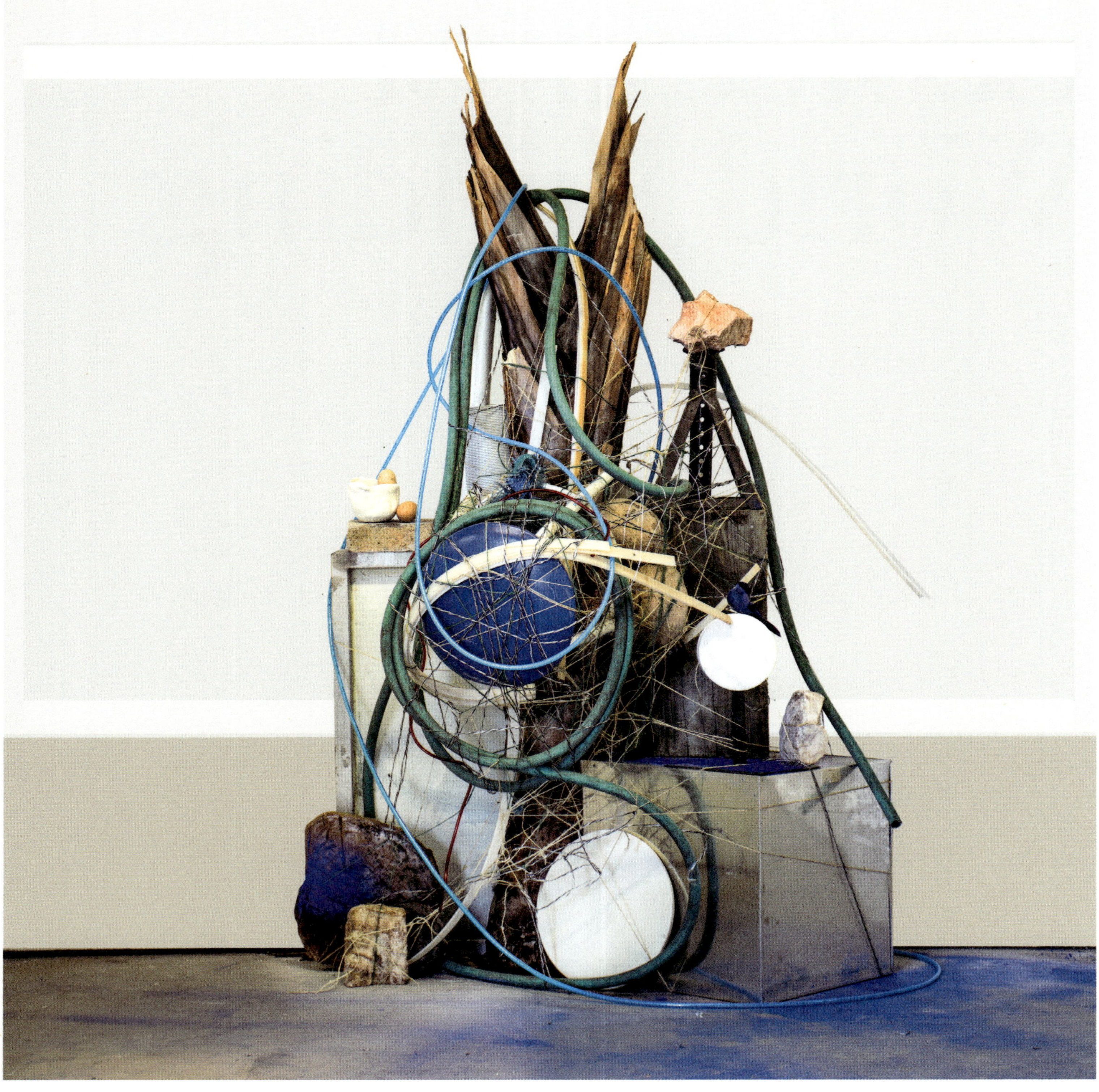

new age remedies drawn from a purified vision of nature—crystals, sound baths, herbal tinctures, and the like—to heal her.

In *Whispering Pines 10*, Cynthia feels overwhelmed by anxiety about climate change and environmental collapse. As Moulton told me, Cynthia "yearns for some way to be practical in the world," yet is paralyzed about how to do so. In the film, as in all of Moulton's work, there is a kind of unboundedness in Cynthia's relation to her surroundings, with objects around her becoming animated as a dancing corps de ballet of hippie paraphernalia. Early in the film, Cynthia awkwardly follows an instructional yoga video, cowed by the grace and flexibility of her TV counterparts. She soon enters a reverie in which a new face appears on the TV screen. It is Julia Butterfly Hill, played by Katie Eastburn, standing before the tree named Luna in which she conducted her two-year vigil. She calls to Cynthia in song while a butterfly painted on her face begins to flit about the screen in a whimsical dance. In a soothing tone, Hill sings a libretto by Hallett drawn from speeches she once gave:

What is your tree? And what I mean by that is I lived in an over one-thousand-year-old ancient redwood tree in California for over two years without touching the ground to keep this redwood tree from being cut down. And when I ask people, What is your tree?, it doesn't mean like, What tree will you climb? but, What it is in your life that calls you to be bigger than what you think is possible for yourself and your world? What is it that calls you to stretch beyond what's comfortable into the places that are uncomfortable and then to realize you are more powerful and magical than your mind could ever have believed?

Cynthia finds herself transported to the redwood forest, and, in a series of fanciful events combined with song, she climbs Luna only to fall, be resurrected, and then die once again when her body is picked apart by cartoon birds. Called to build a better world by Hill, Cynthia in the end feeds her body to animals in a sacrifice that is

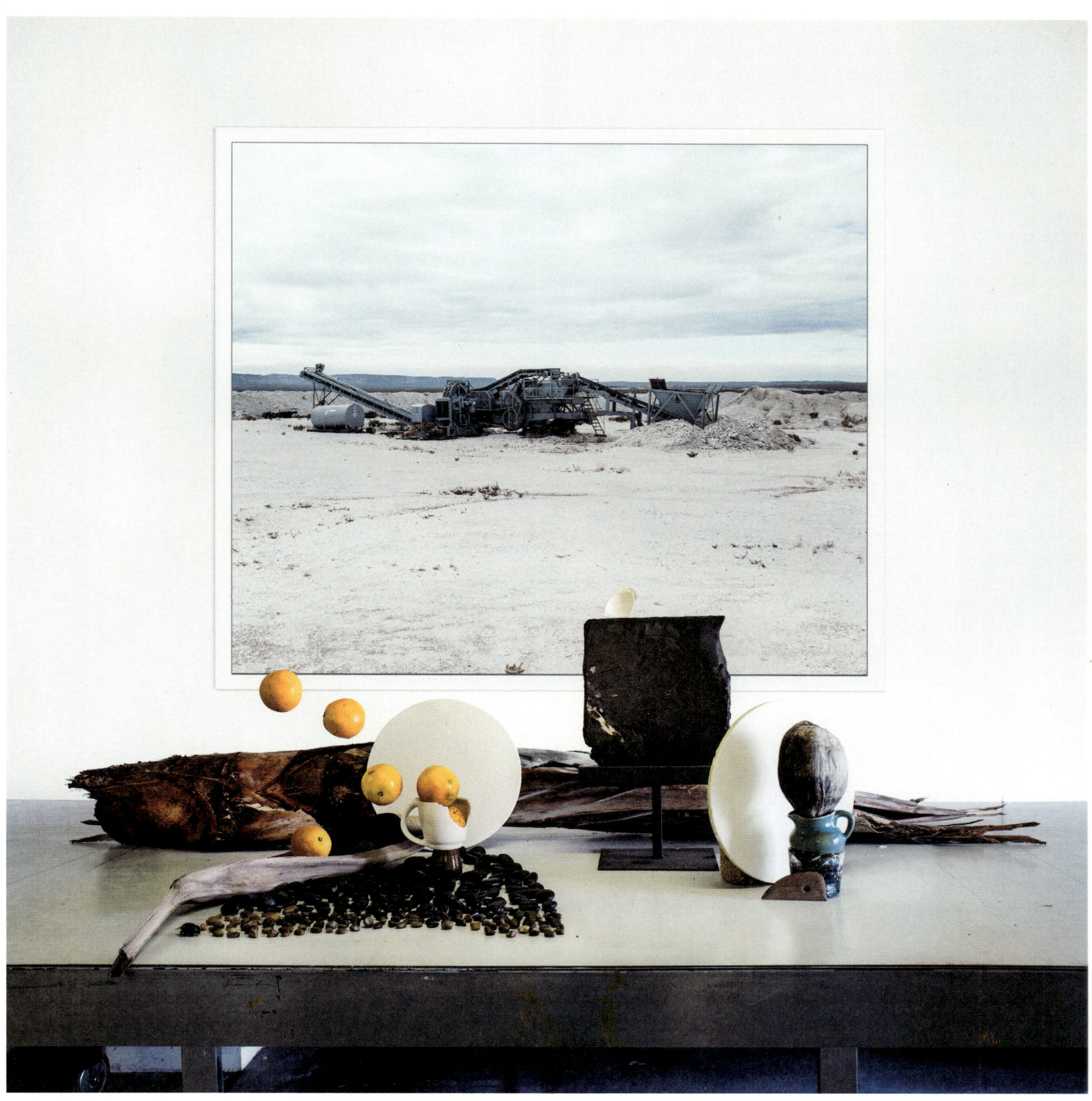

Mary Mattingly's work explores the ways in which raw materials are mined in the name of geoengineering.

both pointless and poignant in its new age–y prostration to a nature she struggles to understand, let alone "save."

In his 2017 book *Facing Gaia: Eight Lectures on the New Climatic Regime*, Bruno Latour notes that one reaction engendered by the loss of mastery discussed by Serres is a push for ever-greater human control of the environment: "And there they are, seized by a new urge for total domination over a nature always perceived as recalcitrant and wild. In the great delirium that they call, modestly, *geo-engineering*, they mean to embrace the Earth as a whole." Yet Latour calls out this compulsion to apply more technology to solve climate change as a fallacy that presumes human authorship of nature: "It is obvious that technological metaphors cannot be applied to the Earth in a lasting way: it was not fabricated; no one maintains it; even if it were a 'space ship'… there would be no pilot. The Earth has a history, but this does not mean that it was conceived."

Mary Mattingly's work explores the ever-more-invasive ways in which raw materials are mined in the name of geoengineering, seeing in this headlong compulsion to extract everything of value from the planet not only greed and poor long-term resource management but veiled geopolitical aims such as military domination. For her project *Because for Now We Still Have Poetry* (2016–18), Mattingly made pilgrimages to far-flung locations in the United States to investigate how photographic materials are mined and produced. Creating photographs about the supply chain of the medium, Mattingly combines these images with artifacts she brings back from often ecologically devastated places to create enigmatic installations. "Real contradictions exist when you're using materials," Mattingly told me recently. "Photography, like writing, can be a form of social justice, but at the same time, there are forms of injustice that go along with its production."

In exploring the extraction of materials that are used for art and photography, as well as for industry, Mattingly wondered why so many minerals are being mined in the U.S. once again, after a period during which cheaper labor pools and less restrictive environmental regulations in other nations were exploited. She came to see links between the opening up for mining of U.S. national monuments, like Bears Ears, in Utah, where she photographed, and the political volatility of nations like the Democratic Republic of the Congo, in which those same materials are also mined. Her research revealed that the U.S. military consumes over 60 percent of the world's high-grade cobalt. It is therefore considered a strategic asset that the U.S. government does not want processed by China Molybdenum, the company that operates one of the world's largest cobalt mines in Congo. Her project took her to a mine in Michigan that also harvests cobalt, as well as a site in Florida that extracts phosphate. In these materials, used for fertilizer, lithium batteries, ceramic glazes, paint pigments, and photography, Mattingly sees a challenge to the notion of image production as creation; perhaps such image making is also a replication of the contradictions of human geoengineering. As Mattingly explained to me in an email:

> Most immediately, photography is a record of a moment that has been able to enter a physical realm; a construction, fiction, fabrication, or truth, it represents what was (seen or unseen). I need it as it is a lens with which I can create worlds. Upon closer examination, photography connects me to complexities and contradictions of a life largely removed from the supply chains that make it up: full of toxicities that I usually do not readily see but may feel the aftermath of, such as its impact on health and the connection photography has to mapping, colonization, militarization, and security. The medium slides precariously in and out of ethical arguments—it can at once illuminate social injustices while simultaneously exaggerating them.

Connie Samaras is likewise interested in the potential for photographic world building, using the archives of one of the great world builders of speculative fiction, Octavia E. Butler, to ground explorations of how to produce work in the era of accelerated climate change. Butler's *Parable of the Sower* (1993) and *Parable of the Talents* (1998) envision life in a near-future Los Angeles devastated by climate change and made a hell of segregated gated communities pillaged by racist and sexist marauders. In Samaras's ongoing series begun in 2016, *The Past Is Another Planet*, she superimposes Butler's handwritten notes and journals on photographs of the gardens of the Huntington Library, in San Marino, California, where Butler's papers are held, subjecting their manicured landscapes to Butler's speculations about how human violence is connected to abuse of the environment.

For Samaras, Butler's archive presented particular problems of access that in some ways mirrored the author's work on the foreclosure of public space in an imagined dystopian future. Samaras

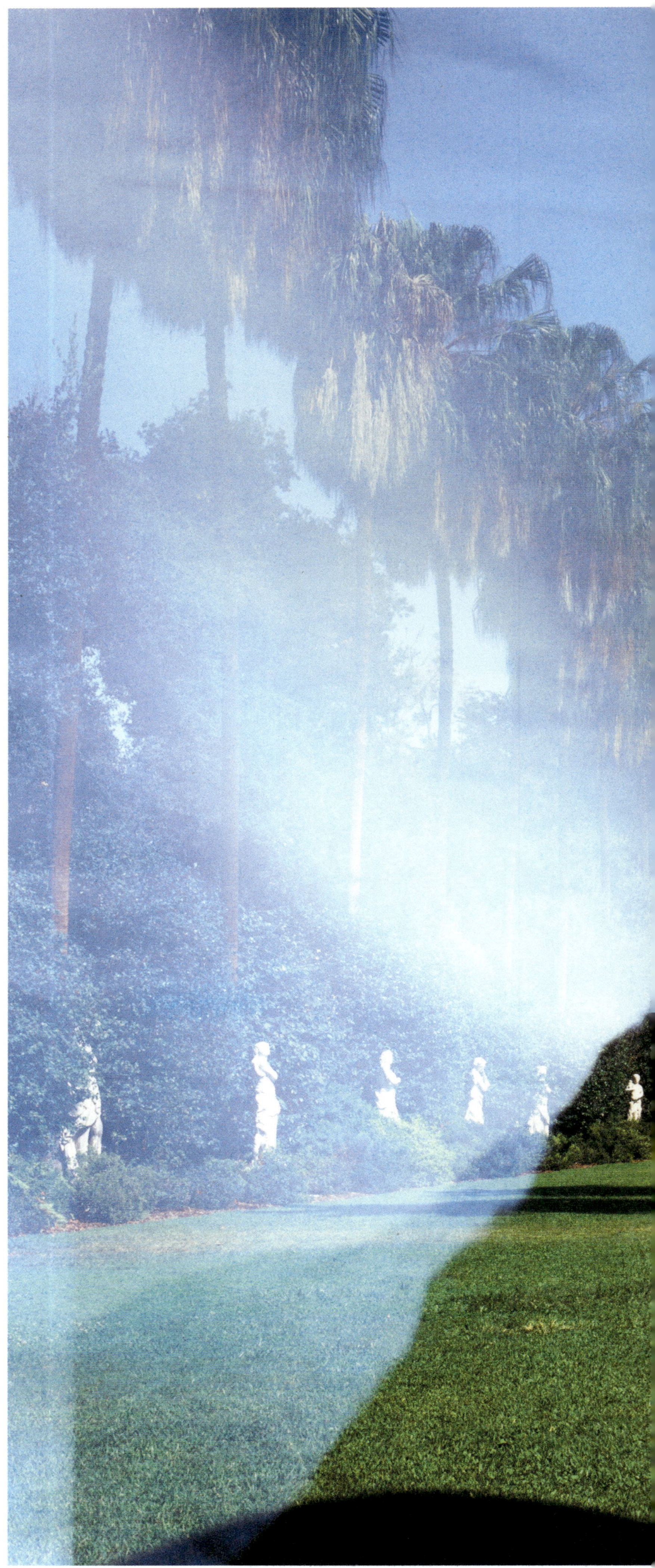

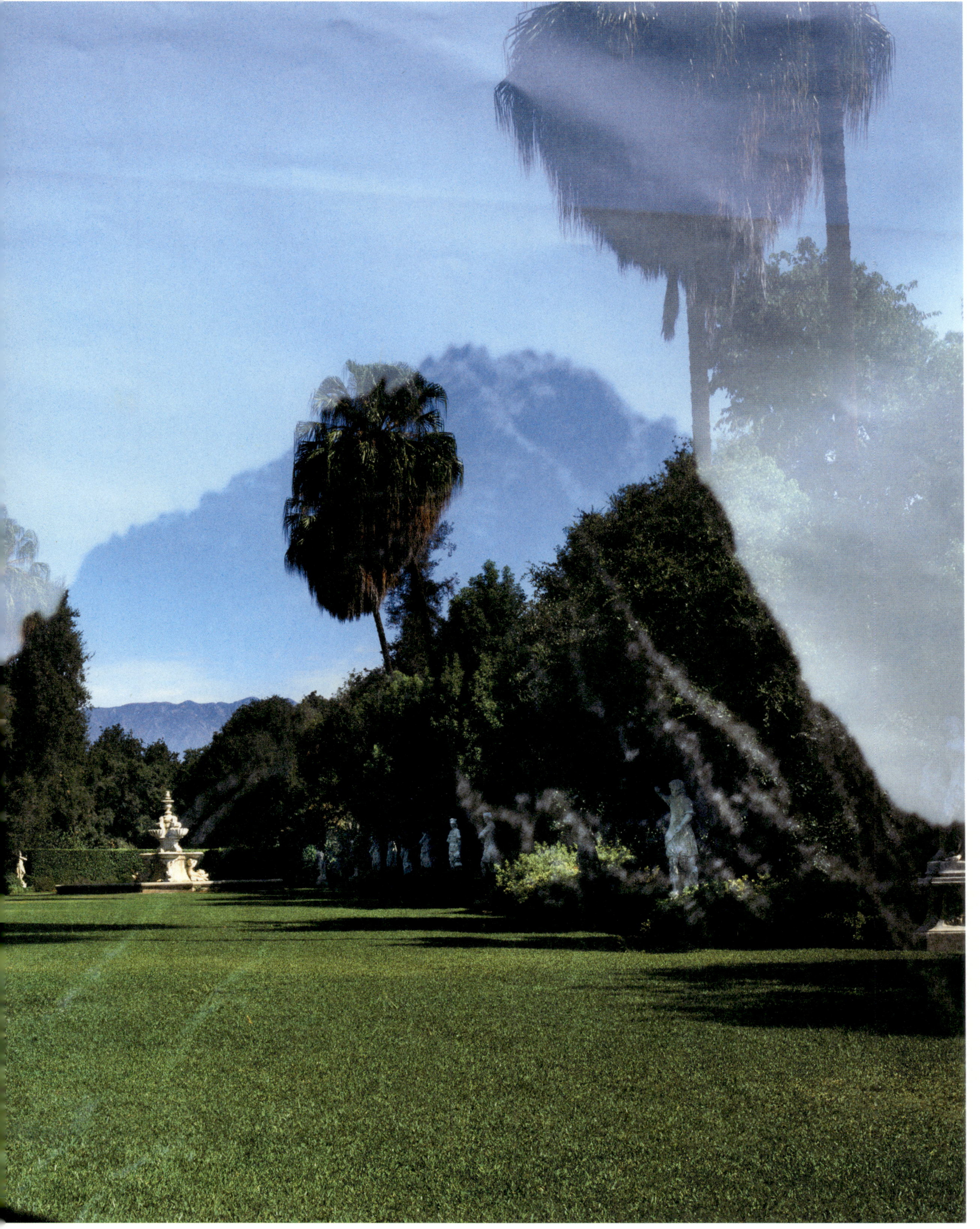

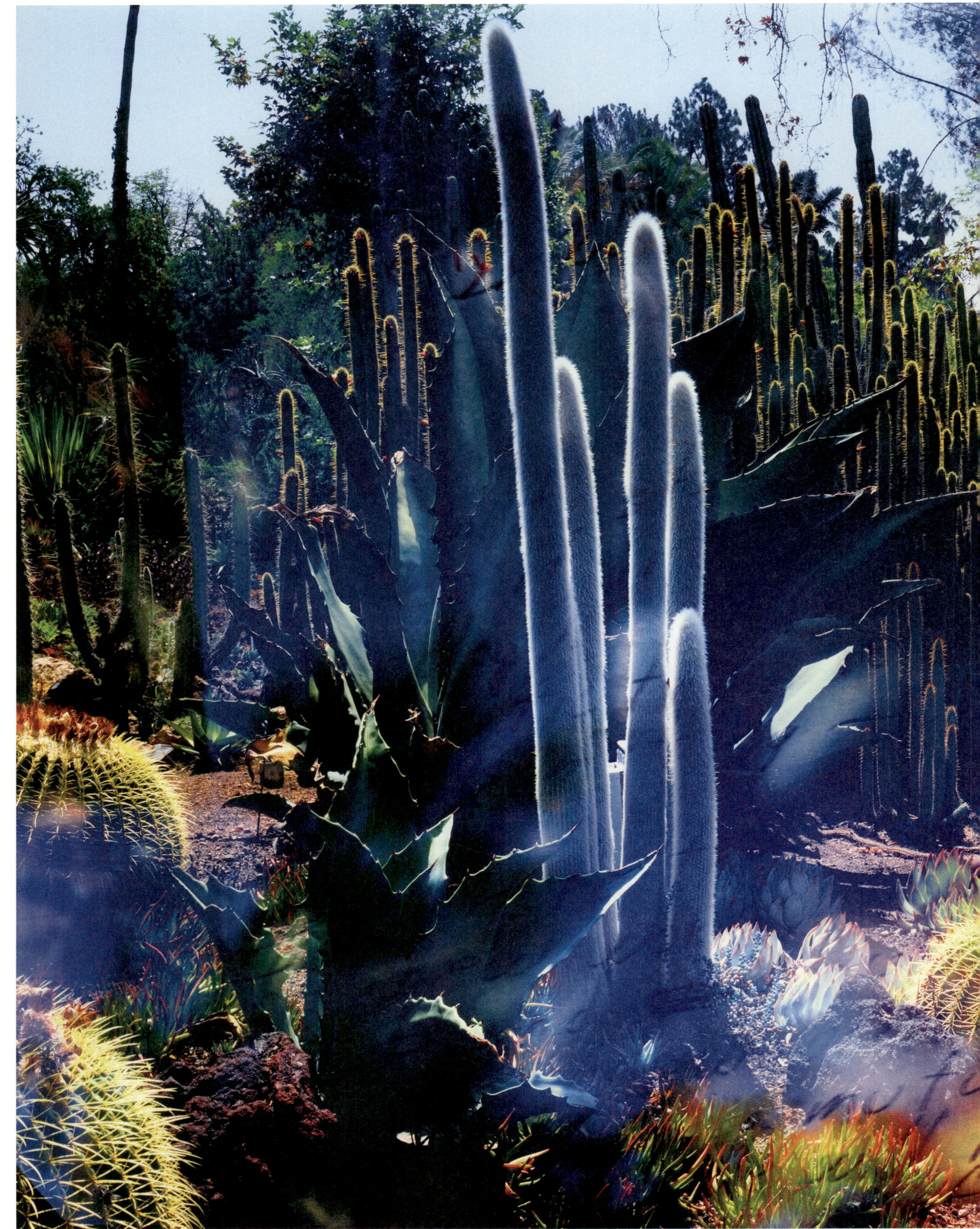

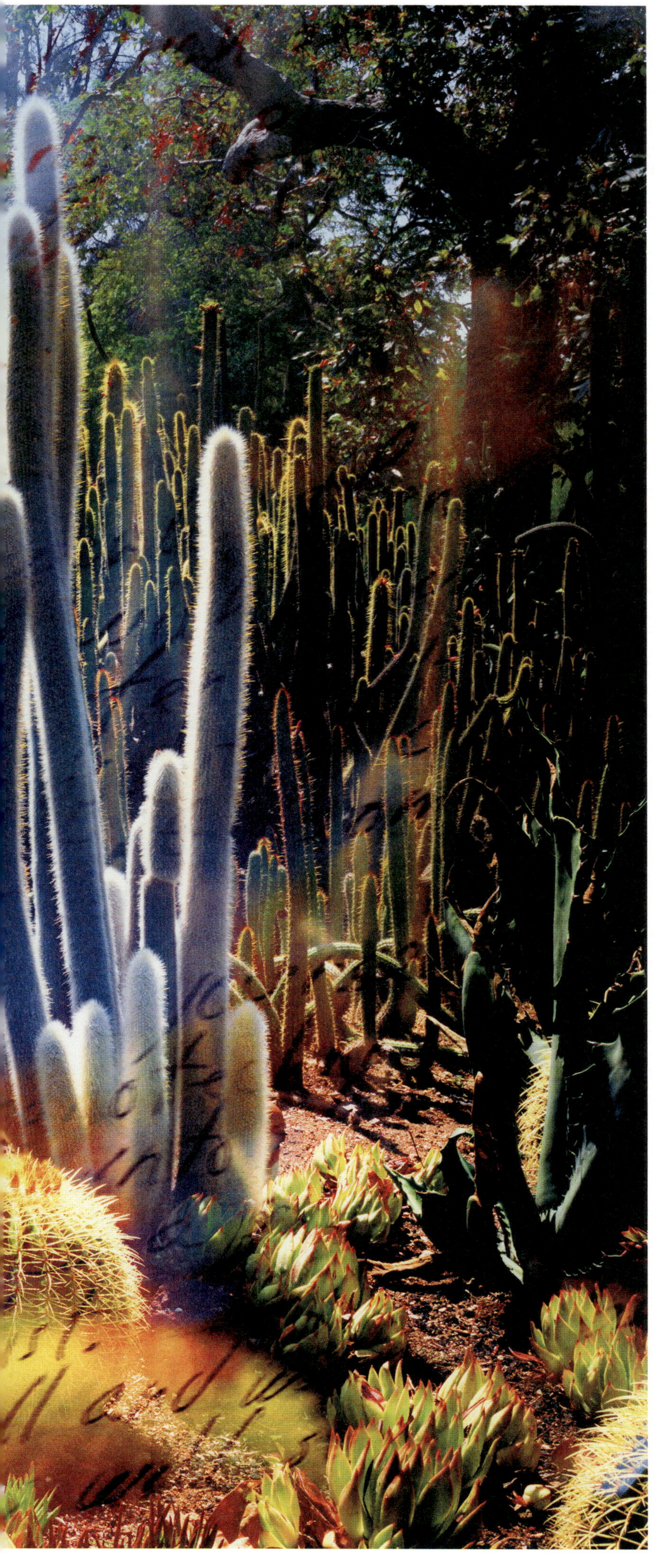

mounted a three-year campaign to be allowed admission to the Huntington Library to see Butler's papers, denied at first because she did not hold a PhD in literature (though she was a tenured professor at a University of California campus). After eventually being granted permission to view the archives, Samaras was also allowed, through her research pass, to visit the renowned gardens surrounding the library when they were not customarily open to visitors, and she eventually began photographing them. These gardens are themselves a kind of human-created no-place of artificiality and unorthodox juxtapositions; for example, a Japanese tea garden, created in the early twentieth century using plants bought from a nearby Japanese restaurant, abuts desert vegetation taken illegally from northern Mexico. Photographing in the early morning, Samaras painstakingly double exposed her images, first shooting the gardens and then holding up Mylar gels with Butler's writings printed on them to capture a palimpsest. (In one work, Samaras overlays an excerpt from Butler's journals that reads, "Los Angeles was dying. Much of the world was changing—changing rapidly, involuntarily blundering through vast climate change.") Because of vicissitudes in the light conditions and glare upon the Mylar, Butler's words become ghostly and fragmentary, interweaving with the surroundings in mysterious ways.

Samaras notes that Butler's papers contain forty years of writings on climate change. While Butler's two *Parable* books portray events that follow from destructive human manipulation of Earth—resource scarcity and environmental damage—they present corporate greed and wealth inequality as the reasons the effects of climate change fall so harshly on the poor and minorities. Though the world building of the novels is often quite dire, Samaras says that she was drawn to the "gritty optimism of Butler, a vision not necessarily about the future, but holding a different kind of vision that you're told you can't have."

Like Serres and Latour, in her recent book *Staying with the Trouble: Making Kin in the Chthulucene* (2016), Donna Haraway criticizes the apocalyptic language that is used to justify defeatism, panic, or arrogant irresponsibility against other beings on Earth— "the 'game over, too late' discourse I hear all around me these days, in both expert and popular discourses, in which both technotheocratic geoengineering fixes and wallowing in despair seem to coinfect any possible common imagination." To Samaras, Butler's work is a powerful corrective to this lack of "common imagination," and she sees in even the conflicted beauty of the Huntington Gardens a sense of hope amid uncertainty. "We are in such dire times," she says, "and I guess the question becomes, How does one survive such things? I share Butler's outlook that it's a collective endeavor and not everyone survives. But the point is that it's collective."

Eva Díaz is associate professor, History of Art and Design Department, at Pratt Institute.

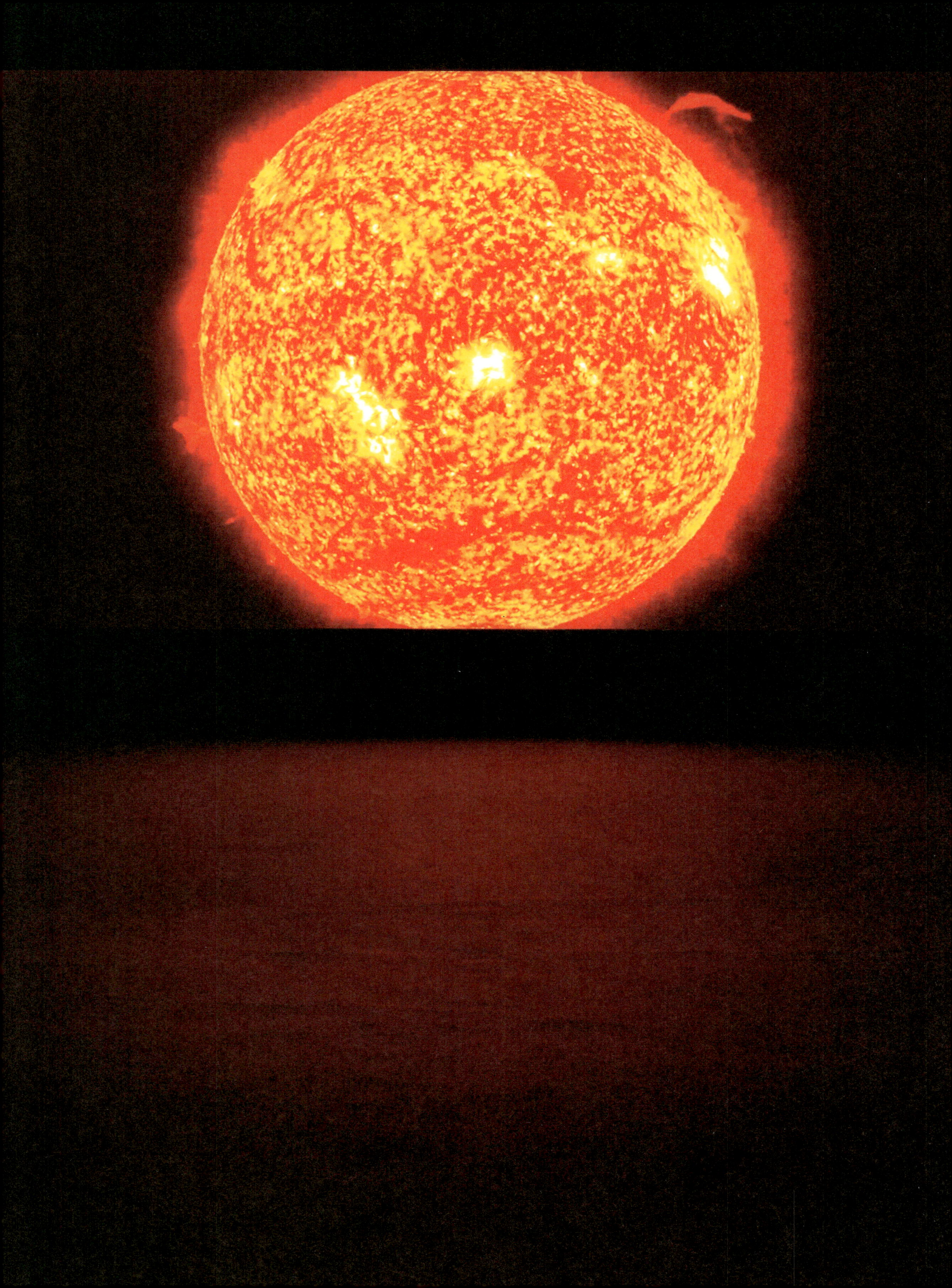

Art in the Anthropocene

T. J. Demos in Conversation with Charlotte Cotton

The word *Anthropocene* is quickly becoming part of everyday language. Describing the epoch already upon us, in which human actions are the determining factor in shaping Earth's geology and ecosystems, the sound of the word alone connotes the gravity of its meaning. Bleak reports of the catastrophic impact of climate change appear with regularity in a moment when politicians blithely deny climate science. Despite current evidence and long-term projections of severe environmental and economic costs, humans are failing to solve a problem of their own creation.

Photography has a long tradition of engagement with the environment, whether providing evidence of exploitation or modeling reverence for the wilderness. In our precarious times, what is the role of visual culture in grappling with a crisis of such magnitude? T. J. Demos has considered the intersections of contemporary art, global politics, and ecology across his many books, which include *Against the Anthropocene: Visual Culture and Environment Today* (2017), *Decolonizing Nature: Contemporary Art and the Politics of Ecology* (2016), and *The Migrant Image: The Art and Politics of Documentary during Global Crisis* (2013). Here, he discusses his work as a writer and thinker and addresses how art making might take a more intersectional approach to visualizing the environment.

Charlotte Cotton: **In the first chapter of your book *Against the Anthropocene*, you raise a question that really struck me: "How does the Anthropocene enter into visuality, and what are its politics of representation?"**

Conversations around independent, authored photography are extremely adept at working through the politics of representation, especially those of identity. But "landscape photography," even photographic works that are at least symbolically aligned with environmental trauma, is rarely positioned under the same critical microscope as other photographic genres. Why do you think we are reluctant to consider the intersectionality of representation and our socioecology?

T. J. Demos: The first thing to remind ourselves is that, as you indicate, there's no unified field of photographic practice, but rather a multiplicity of approaches, many conflictual. Long-standing conventional and dominant ones tend to aestheticize landscapes in ways that exclude conflict and socioecological, political concerns. Landscape has a long art-historical tradition, and the tendency to portray "nature" as a separate realm, defined by the absence of humans and highlighting the beauty of "wilderness," has been endlessly repeated. Yet we know that the construction of landscapes has been part of the colonial project. The translation of that construction into conservation practice is no less predicated upon the forced displacement of Indigenous Peoples and supporting racial and class privileges, something that continues to this day under the aegis of the extractive economy, which also contains a strategic visual component.

In this sense, landscape photography, driven by the art market or commercial journalistic imperatives, tends to support that expansive colonial project, sometimes unintentionally, by practicing the objectification of the nonhuman and its transformation into a commodifiable picture that can be possessed within economies of wealth accumulation. Perhaps some are reluctant to consider this intersectionality because it threatens not only deeply held beliefs and aesthetic values, but also economic interests.

CC: **How has the pronouncement of the Anthropocene affected landscape photography's approach to image making?**

TJD: With the Anthropocene epoch, we're witnessing a shift in visuality toward postphotographic remote-sensing, where the landscape becomes regionalized, becomes the Earthscape. The image is not only directed toward commercial markets, but also toward the technoscientific corporate-state-military complex, in the name of surveillance, climate data modeling, green capitalist rationality, and geoengineering.

The problem here is that the environment is once again reified as a discrete realm, cut off from sociopolitical realities. Environmentalist activism often follows suit by challenging carbon pollution but also accepting the delimitation of what *climate* means. By doing so, it perpetuates the nature-culture divide and limits its own intervention in the science that is alienating and irrelevant to the present urgencies of many submerged in the conditions of everyday state, corporate, and police violence.

Meanwhile, genres of portraiture and social documentary, for their part, tend to reify their own respective categories, failing to consider how present climate transformation exacerbates economic inequalities and social violence.

CC: **What alternative approach are you calling for?**

TJD: One thing I'm calling for is the disarticulation of the term *environment* into its many possible meanings so that we can

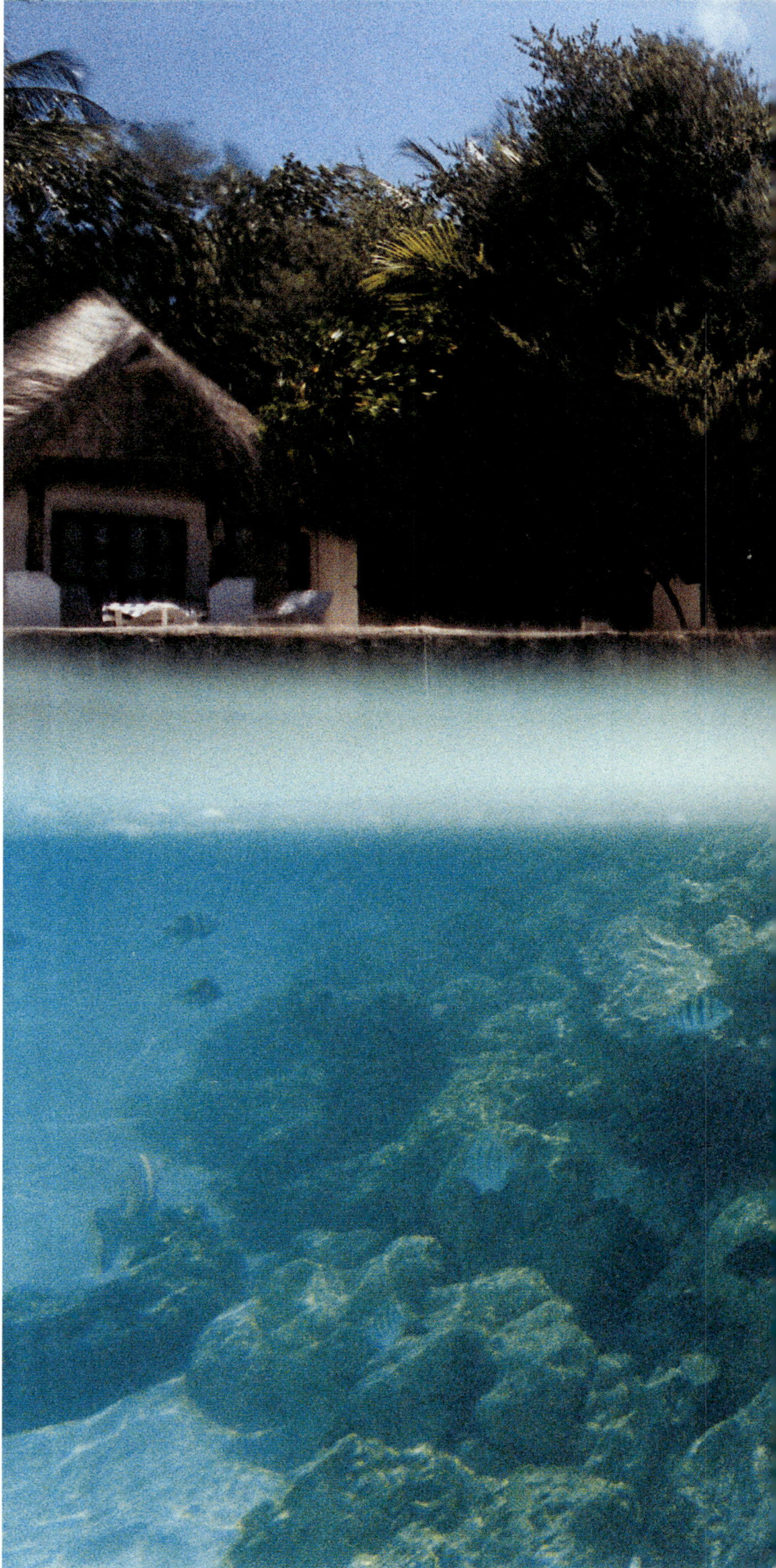

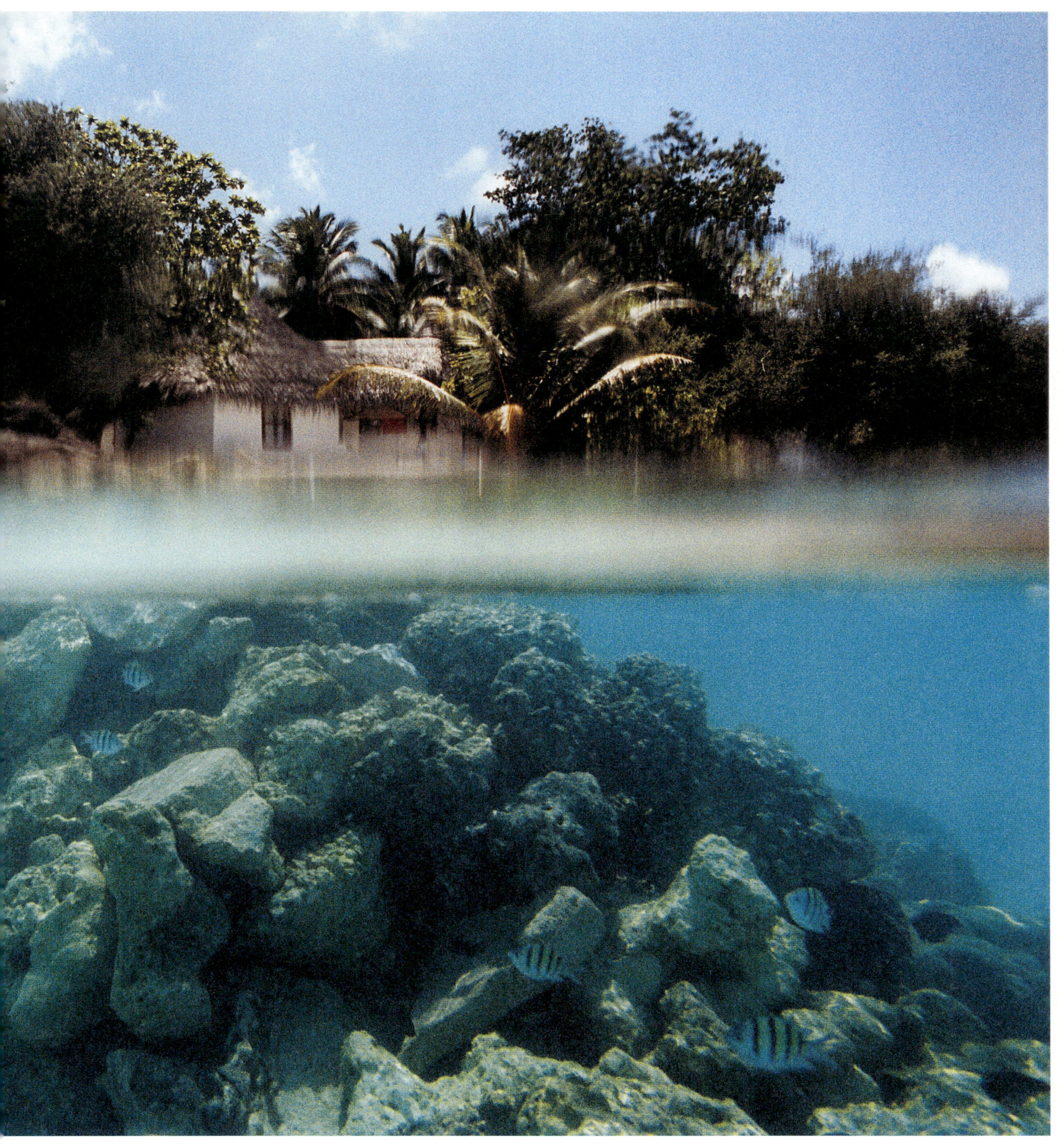

I'm interested in exploring these convergences between political force fields and aesthetic emergences.

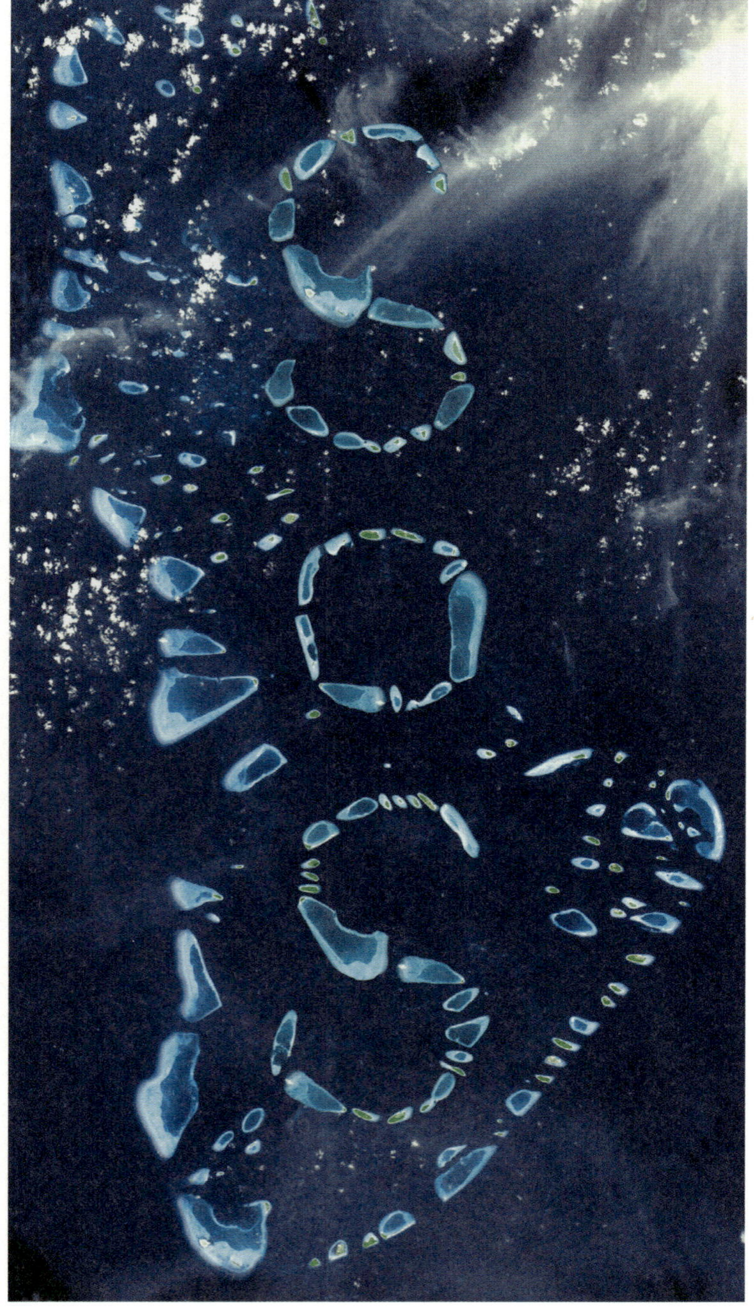

recognize and engage with cross-sectoral conditions. A "climate" might be one of antiblackness (as Christina Sharpe writes, as John Akomfrah and Arthur Jafa visualize). "I can't breathe" is not only a matter of police brutality directed disproportionately at people of color, but also a matter of polluted air owing to the Capitalocene, where geology is increasingly determined on a global scale by our economic order, and its violences and inequalities. I write now from a burning California where it's unsafe to be outside for extended periods—but for the multitudes who are houseless, there is no option.

An intersectional approach would insist on seeing the visual field as structured by these inextricable relations of power, economic forces, and ideological mechanisms. Certainly there are numerous practices today attempting to do just that. Works by Forensic Architecture, Ursula Biemann and Paulo Tavares, Laura Kurgan, and Richard Misrach in collaboration with Kate Orff and Scape, to mention only a few, are exemplary for me. Such an approach might also include focusing on sites of environmental trauma, in order to raise awareness or inspire new legal orders based in biocentric imperatives. Yet even here there's a danger—that of aestheticizing destruction, something I address in *Against the Anthropocene*. For example, the epic photography of Edward Burtynsky, for me, calls up Walter Benjamin's Nazi-era but still resonant critique of a political aesthetics that relishes scenes of self-destruction—which is not helped by Burtynsky's determinedly apolitical self-positioning and market-directed practice.

CC: **You've mentioned a number of contemporary art practices you admire. One of the elements I appreciate so readily about your writing and thinking is that the artists you focus our attention upon—which include Josephine Starrs and Leon Cmielewski, the Argos Collective, Amy Balkin, Ravi Agarwal, Kristina Buch, and the Otolith Group—are revealed to the reader rather than offered up as illustrators of a theory.**

TJD: In creating images, framing points of view, arranging affective sensation, and reconfiguring perception, artworks exhibit intelligence, model forms of life, produce subjectivities, and enact politics. In my work, I'm always interested in exploring these convergences between theoretical writing, political force fields, and aesthetic emergences, where art plays an active role in constructing intersections.

When I look at the work of the artists you mention, my ultimate objective is to get at that distinctive movement that only this or that particular work achieves. I try to honor its contribution by thinking with it, and by articulating the resonances that speak to the relevance and significance of its project.

CC: **In your perspective, where and how do artists shape our socioecological narrative?**

TJD: Without perpetuating the notion of the heroic, exceptionalist quality of art that's long been part of the avant-garde mythos, I do believe that art is able to shape narratives in unique ways. Though art history and criticism have been, as art has, corrupted by markets, they still hold the potential to redeem art as a place where we can invent, experiment with, deliberate, and critically consider emergent forms of life, which is more urgent than ever, now that we're facing an ever-more-likely near future of mutually assured self-destruction.

This points to the sociopolitical and, indeed, ecological significance of artistic practice as a laboratory where we can create, restore, and decolonize futures on the basis of social justice and multispecies flourishing, where social transformation can be

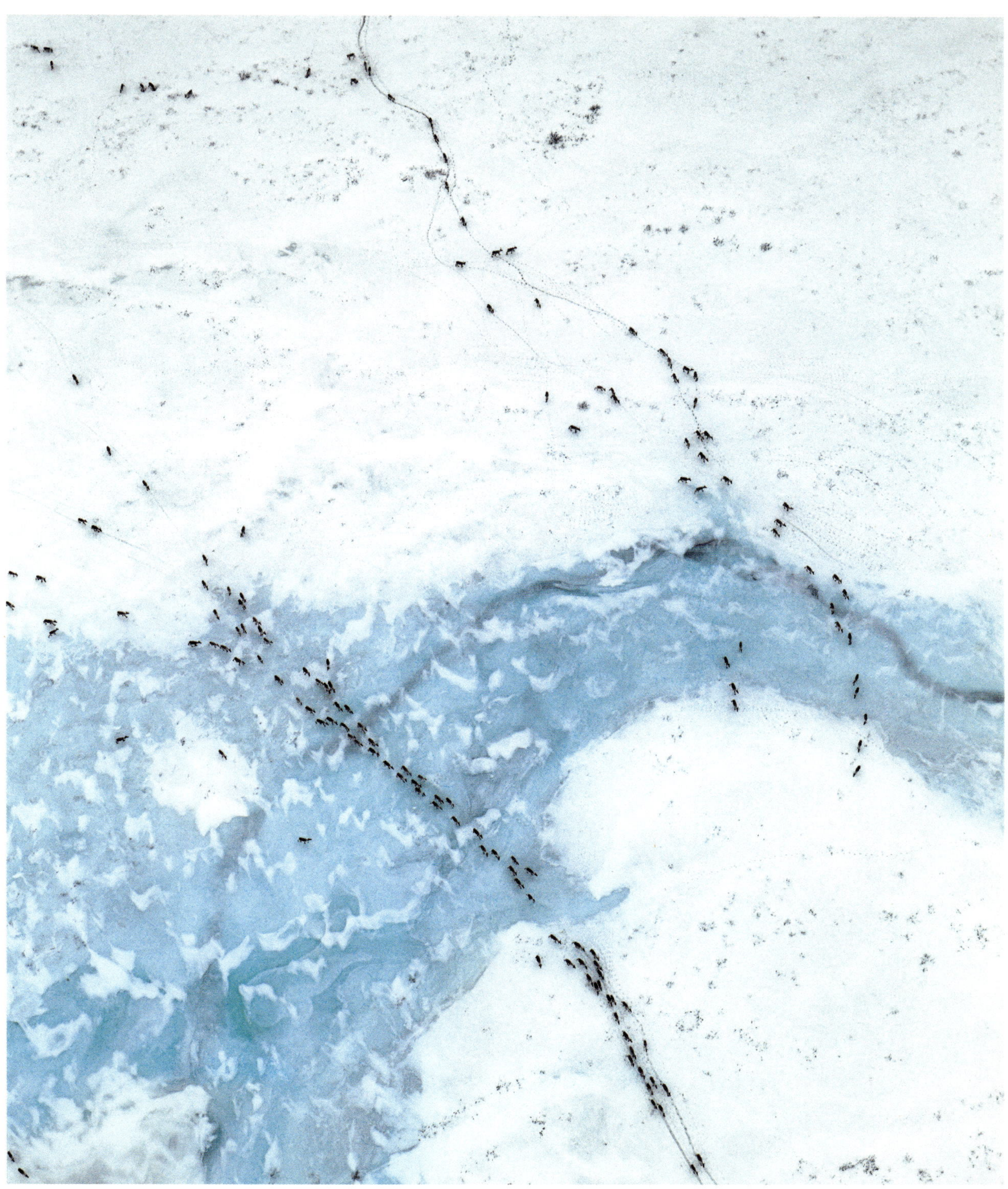

advanced, where we can "stay with the trouble," as Donna Haraway advises. It's a place where we can insist on the importance of anti–anti-utopian thinking—thinking against the nihilism and cynicism that otherwise rule the current hegemony of capitalist realism.

CC: Can you illustrate how art might create meaningful space for this kind of thinking?

TJD: One of my most recent essays is on Arthur Jafa. Looking at Jafa's work, in particular his video *Love Is the Message, the Message Is Death* (2016), allowed me to open a dialogue between environmental studies and its technoscientific leanings, on the one hand, and social-justice critiques of racial capitalism, on the other. By situating this conversation alongside Jafa's video, we can avoid what some call white environmentalism, or ecologies of affluence—modes of advocacy based on privilege that seek to sustain livability without addressing profound social inequalities—while also pushing antiracist activism toward wider considerations of unjust atmospherics and ecologies of inequality. Ultimately, the art allows us to think with it in the experimental formulation of new collectivities that might actually contribute to widening social transformation in crucial and necessary ways.

CC: **You write about how artists can provide us with proximity to our socioecology, and, therefore, to some hope of social transformation. I am curious where your own proximity to our socioecology stems from.**

TJD: I first had the chance to address political ecology in a catalog essay for *Radical Nature*, an exhibition at the Barbican in London in 2009. I wrote about the ideological functions of sustainability discourse in environmental art and activism, where, as it turns out, "sustainable development" has always meant the imperative to sustain economic growth before all else. Meanwhile, I had been researching politico-economic conflicts under globalization since

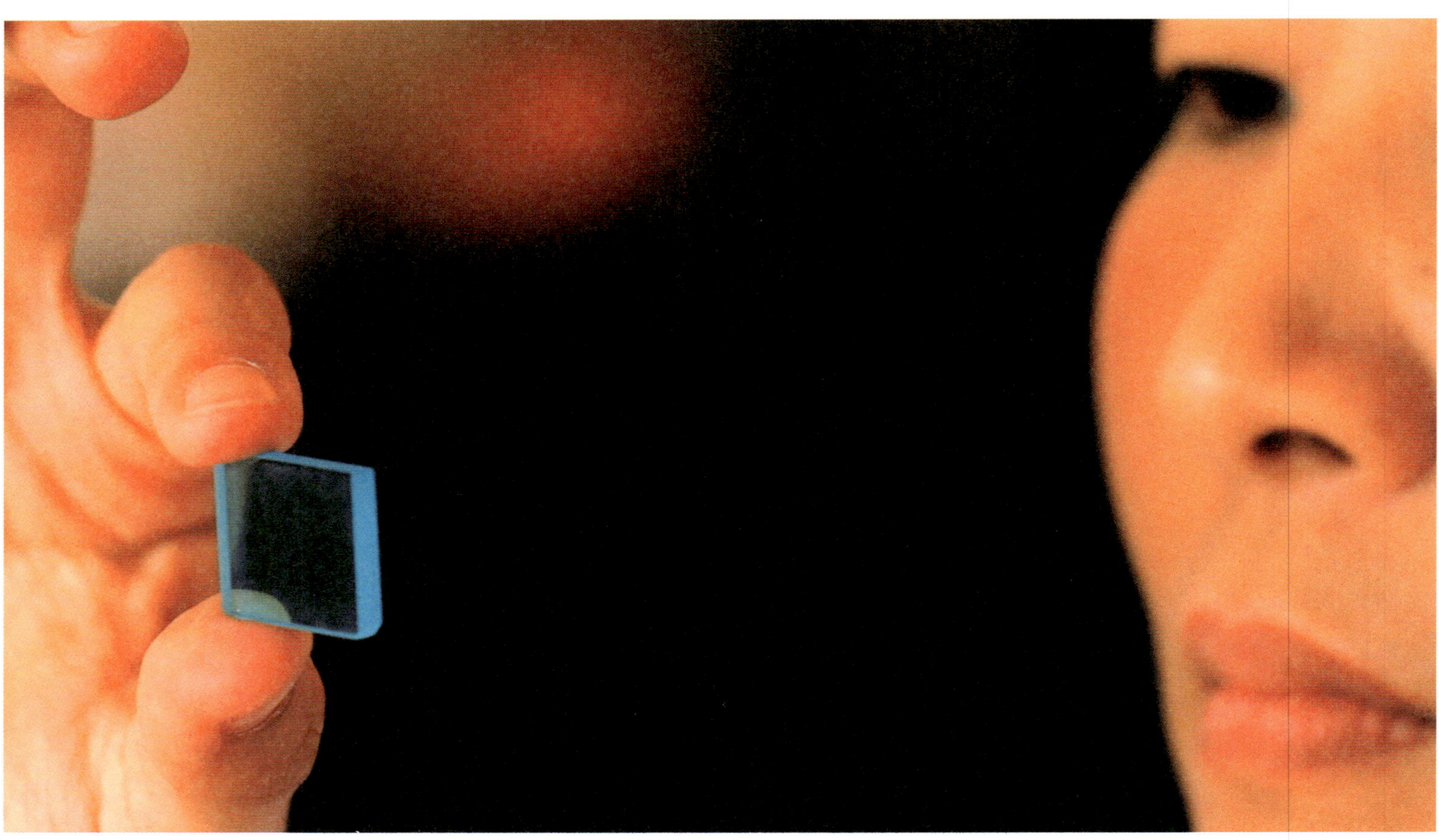

It not only matters *that* we address this crisis, but *how* we do so.

1989, particularly in relation to U.S. military zones, migration and border control, and the way many artists were investigating these subjects, which led to my book *The Migrant Image*. It was only a logical step to consider the environmental impacts of our world economic order in turn.

Soon it became clear that environmental violence was not simply a peripheral problem to social inequality and state violence, but integral to globalization. What's more, conditions were gradually worsening to the point where our very livability as a global civilization was increasingly seeming imminently at risk. What drives my work, after years of researching ecology, stemsfrom the basic activist imperative I feel, which requires doing everything possible to contribute to the movement to stop catastrophic climate breakdown, and to work toward solutions grounded in social justice rather than green capitalism. It not only matters *that* we address this crisis, but *how* we do so, and it's clear that financial elites, for instance, are already mobilizing climate-change responses to serve their own interests.

This, as journalist Allan Nairn points out, is allied with "incipient fascism" in the U.S., mobilizing the worst elements of white supremacy and antimigrant xenophobia to reach its goals. We're facing a war of the worlds, and we must do whatever we can, as well as all we can, by advancing a progressive and intersectional agenda.

CC: Is your commitment to writing and teaching driven by a desire to serve the human imperative?

TJD: Writing is a key instrument for me, and it connects to researching, collaboration, teaching, and activism. Like art, writing isn't illustrative or supplemental to thinking or meaning making. It's a generative process. Through its very difficulties and revisions, mistakes and corrections, dead-ends and breakthroughs, it allows and provides the material conditions for new insights and realizations to emerge, for positions to be tested and taken, for commitments and political stakes to be articulated.

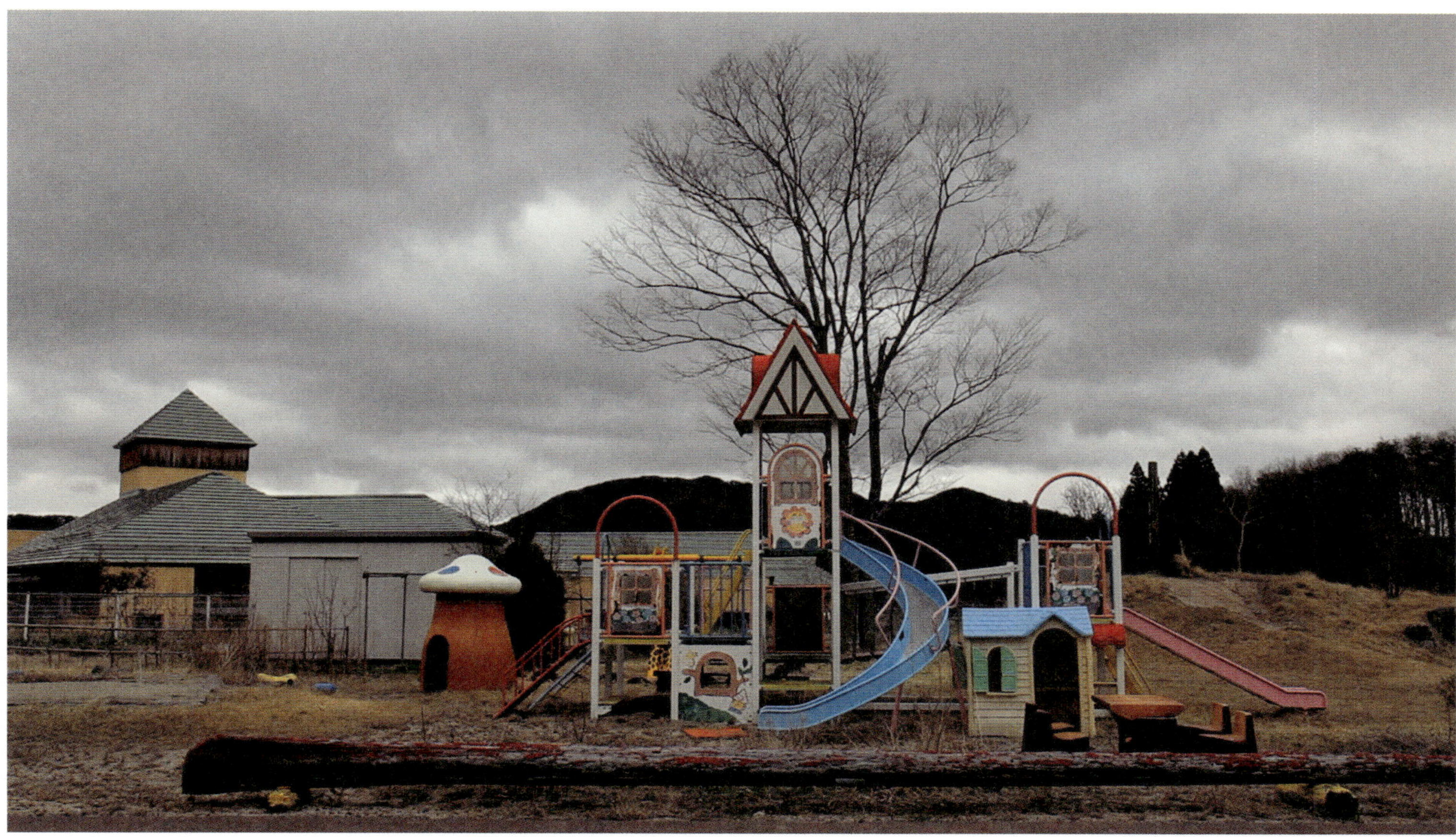

That said, I don't generally speculate about where my texts might end up someday, or how they'll be regarded in the future. Certainly we can think of books as messages to the future, as time-travel machines, and I definitely consider past literature in this way. Take experimental sci-fi where the text is a place where time-travel can occur, as in Octavia E. Butler's *Kindred* (1979), or the Otolith Group's notion of erstwhile events as holding within them past-potential futures, which might be critically decoded and newly mobilized in the present—also part of the magic potential of photography, you might say.

In my recent writing, though, I'm more interested in writing as a site where we can collect and reflect on messages from the future by considering multiple, conflictual potential movements that are now at stake. Knowing that things can get worse, even to the point of the end of human civilization as we know it, ultimately drives my work. I figure it as a contribution to social transformation, which nonetheless, as I'm well aware, may still not be enough to save us.

Charlotte Cotton, a writer and curator based in Los Angeles, is the editor of the Aperture books *Public, Private, Secret: On Photography and the Configuration of the Self* (2018) and *Photography Is Magic* (2015).

Paradise &
Dystopia

Thomas Struth in Conversation with Aaron Schuman

For more than three decades, the German artist Thomas Struth has made photographs of startling clarity and precision. He approaches his subjects—New York streetscapes and South Korean skylines, German families and Queen Elizabeth II, the Louvre and Disneyland—with the objective eye of a journalist and the meticulous composition of a painter. Associated with the Düsseldorf School of Photography, which emerged in Germany, in the 1970s, under the teaching of Bernd and Hilla Becher, Struth's work is also connected to the psychological depth of August Sander's portraits and the conceptual rigor found in the paintings of Gerhard Richter, his former mentor. Struth's recent book *Nature & Politics* (2016) extends his fascination with natural landscapes into a critique of human engagement with the planet, looking at the built environment.

On an uncharacteristically warm day last October, the photographer and writer Aaron Schuman visited Struth at his second-floor Berlin studio, overlooking the leafy banks and calm waters of the Spree River. Several studio assistants worked at their respective monitors; one table revealed scale models of art institutions around the world—including MAST Foundation, in Bologna, Italy, where Struth's touring exhibition *Nature & Politics* opened earlier this year. Together, Schuman and Struth discussed technology, animals, and how climate change might open the door to collective effort.

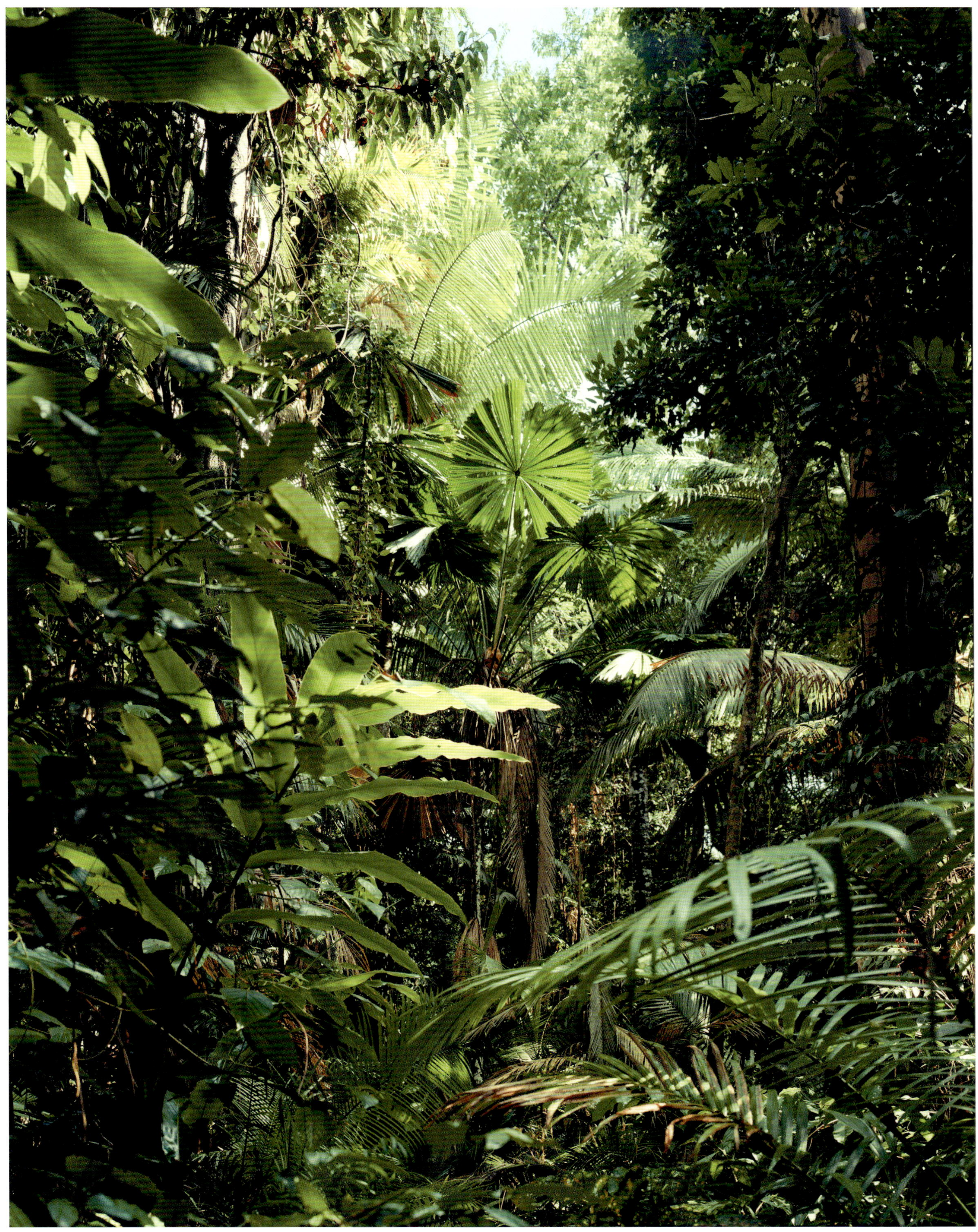

The connection between the "goal" or the "target" and yourself was fascinating. I thought about the camera as the ultimate pointing of the bow.

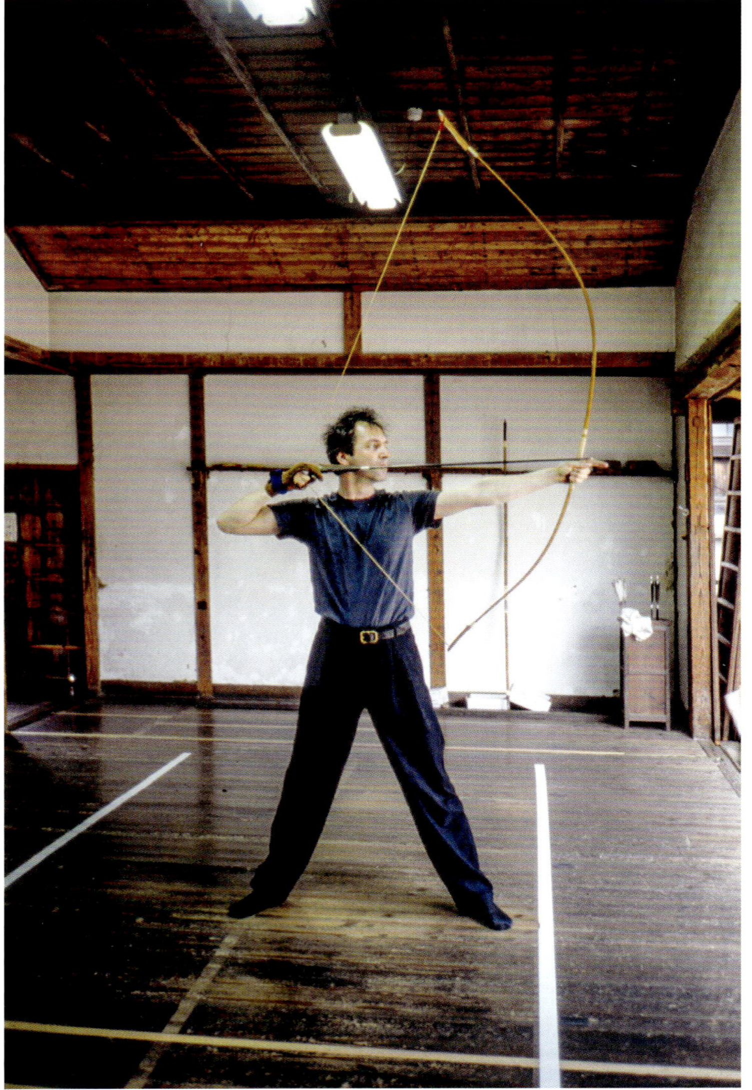

Thomas Struth: I'm sorry I'm a little bit late—I've just come from a funeral.

Aaron Schuman: **I'm sorry to hear that. Were you close to the person?**

TS: She was a neighbor who lived in our building, upstairs. She'd just turned fifty, and was the mother of two kids. She was East German—very sporty, very active, and was kind of a life force. My experiences with many people who were born in East Germany is that they are very natural in a way, in their whole demeanor, because they didn't grow up with advertising and these glossy role models that put psychological pressure on them. It's completely different from capitalist societies. They have this kind of joie de vivre, because the main thing that they enjoy is other people—being connected. At the funeral, there were maybe three hundred people or so.

AS: **It's interesting that you used the word *natural* to describe the East German mentality. That comment seems to reflect something about your psychological perspective, your photography in general, and, more particularly, how you scrutinize the world in your recent body of work, *Nature & Politics* (2008–13).**

TS: There are many categories of influence within human existence that could be scrutinized—there's the physical, the social, the political, the psychological, and so on. In a sense, the psychological field is difficult to read or to treat within the category of science—there's a desire to approach it from a scientific perspective, and it's much studied in this way, but still it's not so clear. Nevertheless, it's important. When you look at people like Donald Trump, or Recep Tayyip Erdoğan, or [the far-right German politician] Alexander Gauland, you can't help but think, What's going on inside them psychologically? How can they walk the path that they're walking on? In a way, it's like dark matter. It is dark matter.

AS: **Your works have often been read and interpreted from political, environmental, and sociological perspectives, but do you feel that this psychological component is just as vital in terms of the way you work?**

TS: Yes. I believe that psychology comes into it when I consider things like: What am I attracted to? What's my subject matter? What's my longing? What do I choose to evaluate? That's very important to me—what you could also label as intuition. When I read art—for example, when I look at Gerhard Richter's abstract paintings or a Mark Rothko—I partly understand them as a psychological network. Psychology is a big component in terms of evaluating my perception of what I'm looking at, and also in terms of what I want the viewer to see.

AS: **How do you initially determine what to photograph, and then how do you decide how to photograph it?**

TS: When I started to photograph, at the age of twenty or so, I was still painting, and I chose to photograph things that had a lot to do with me. I was a student, walking to and from the Kunstakademie Düsseldorf a lot, and I was lonely. I'd just left home and was looking for new friends, a new environment, and a place in the world. This was in postwar Germany, so I was walking in this scattered environment that was very heterogeneous—it wasn't like Paris or New York.

AS: **There wasn't a predetermined logic to the environment, in terms of how it was organized.**

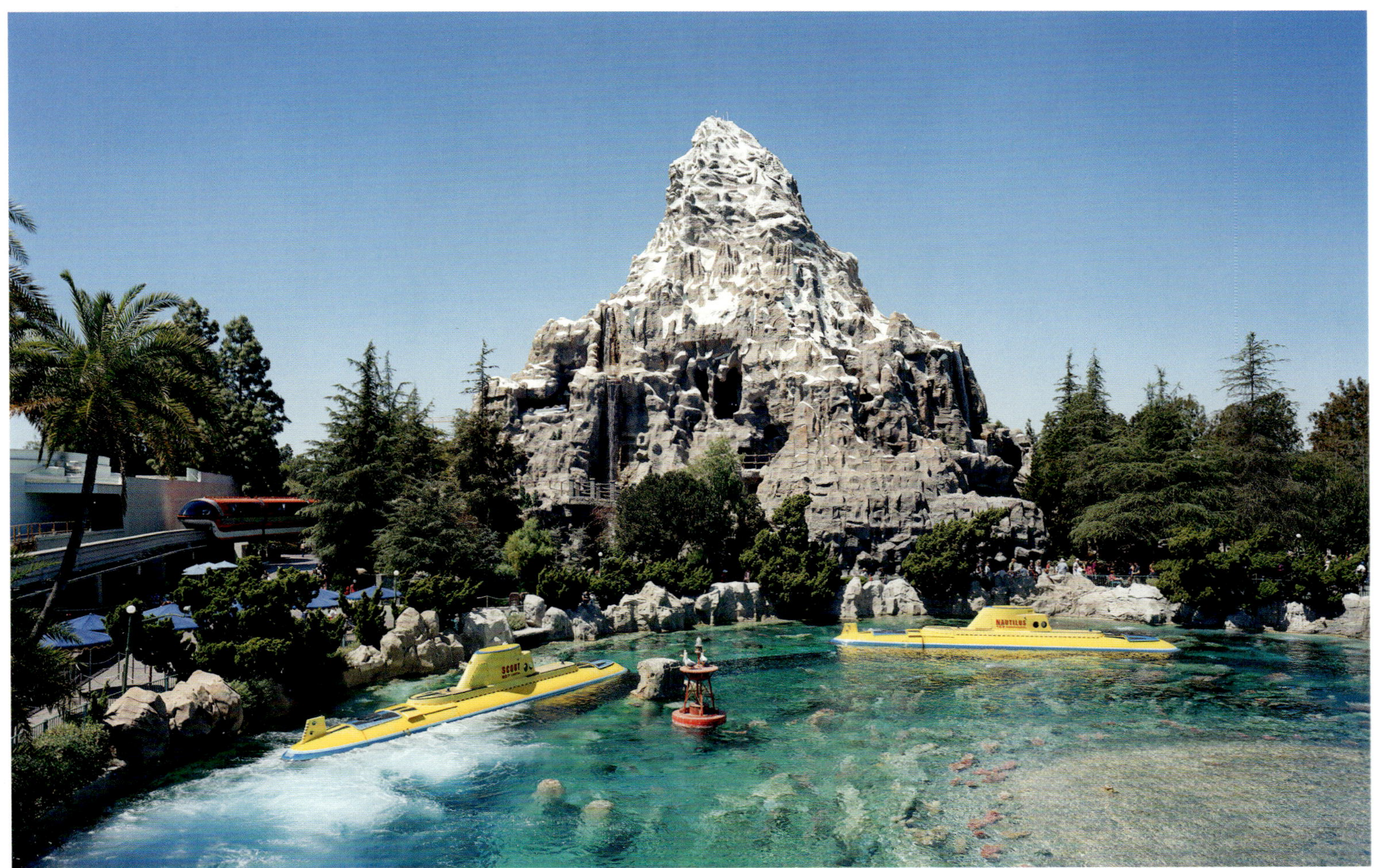

TS: No, there wasn't a determined logic, so I was living in this sort of ruptured place. For my generation, there was a strong focus on the individual, and the idea that you can transform yourself through life and experiences. But because the environment was ruptured, you couldn't find consistency, you couldn't escape yourself, and you couldn't be sentimental, so it offered the opportunity to look at the world in a more objectifying manner. The quest and challenge was always to figure out what is really closest to me, or what comes up from inside me, and then try to look at and experiment with it in relation to the outside world in an unsentimental way.

For example, the apartment I was living in was on an elevated ground floor; it had a terrace in the back that looked into the communal space, lots of trees, everybody's gardens, and so on. I often sat outside—eating breakfast or dinner, or just reading—and I would look into the structure of the branches and the trees. I began to think it would be interesting to photograph that, to have a picture plane with a lot of structures within it. At the time, it was just an intuitive curiosity and developed from something emotional.

But years later, I decided to photograph forests and jungles for a series called *Paradise* (1998–2007). I had several previously scheduled trips on my agenda, including one to Australia, where I was participating in the 1998 Biennale of Sydney. I looked at the map and saw that there was a jungle in the northeast of the country, in Daintree, Queensland, so I started there. Then I had a trip to Japan, and another to China, and each time I made sure to go to a forest or jungle to photograph.

So that's one body of work that started intuitively, without me really knowing at the beginning why I was doing it. And to be honest, at first I was worried that people might think that it was just jungle wallpaper. But after I returned from Australia, I made a big print and hung it on the wall of my apartment, and I really liked it—still not knowing why, not really knowing how I got there, but just deciding intuitively that I liked it.

AS: In retrospect, do you think that your positive response to that print was based on its aesthetics, or was it something more emotional or psychological?

TS: It was both. I liked the fascination of identifying certain compositions in an environment where millions of compositions are possible, 360 degrees around me at any one point. And then, when I started to show these works in galleries, I hung several of them in one room, on every wall around. I realized that the effect it had was very quieting—almost meditative. The more I thought about it and showed it in this manner, the more I realized that this work was about nonidentification—just being.

There was a period of time, in the 1990s, when I was experimenting with different kinds of therapy, trying different types of meditation, and doing a lot of "journeys into the self." I went to a therapist in Wiesbaden who worked a lot with breathing, and it became about being in a room without doing anything—just silently being. Also, when I was in my early thirties—maybe in 1986—through a connection with close friends, I started to practice tai chi with a Chinese grand master, which I did for about twelve years. He died about two years ago. In a way, that is also somehow connected, because when you do tai chi, it's a set of very slow movements that don't have an immediate purpose. But when you do something very slow, you become extremely aware of every position, and of the connection between your own body, your mind, and the space around you.

AS: Would you describe your photographic process in a similar way?

TS: Practicing tai chi for a long time definitely sharpened my perception in certain ways, because I became very aware of being in and moving through space; it helped me become more alert to very minute differences.

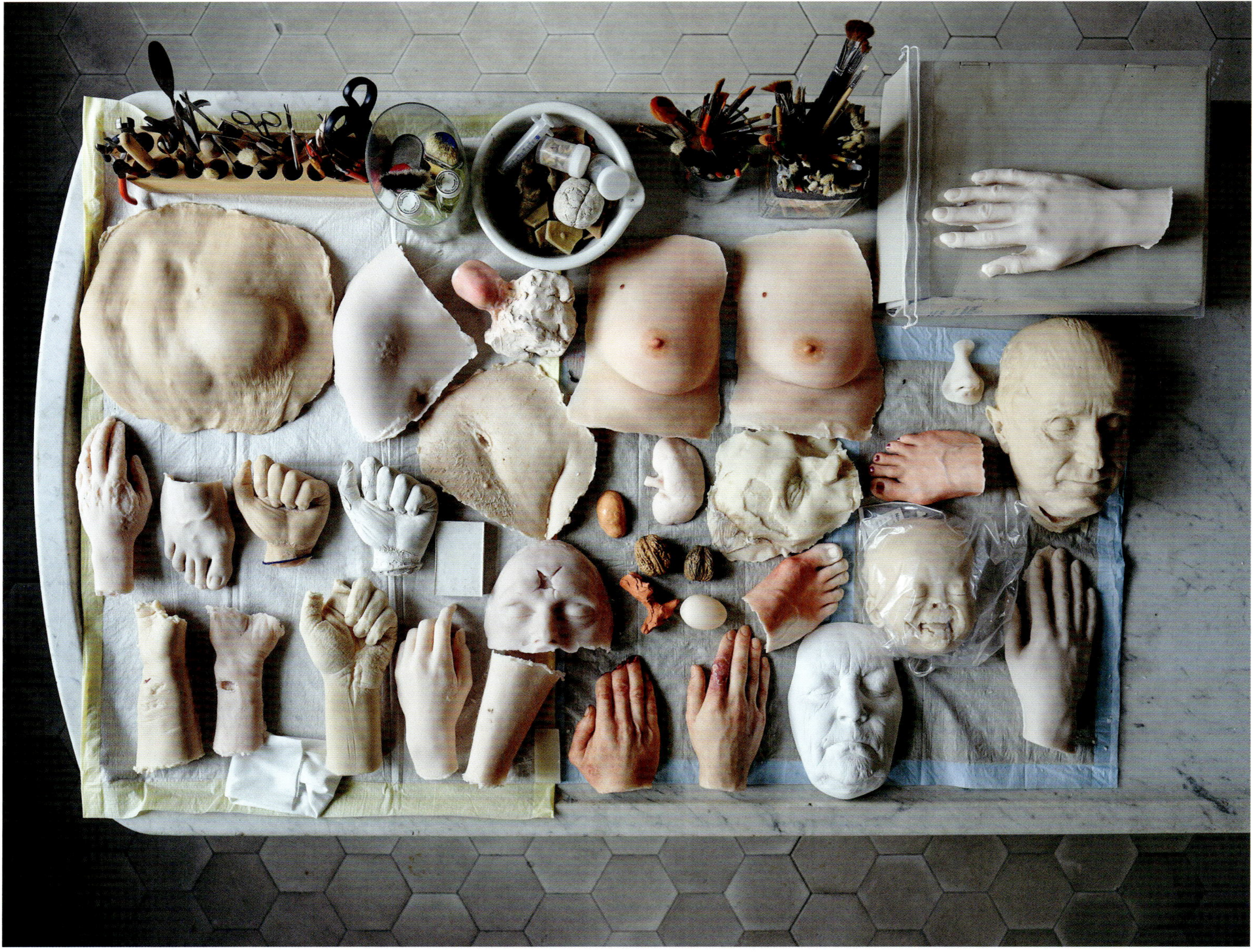

Also, on one of my trips to Japan, in the 1980s, I said to my host, "I'd be interested in learning more about archery." I wasn't really being that serious, but before I knew it I had an appointment with a teacher, and, for a week, I spent every day with him, from morning to evening, learning about Japanese archery. Eventually they asked me, "Do you know of the German philosophy professor Eugen Herrigel?" He was in Japan in the 1920s, taught philosophy, practiced archery, and wrote this book called *Zen in the Art of Archery* (1948). At the time, I hadn't heard of him; I bought his book, and since then I've given it to at least fifteen or twenty people. The connection between the "goal" or the "target" and yourself was fascinating to me; somehow I thought about the camera as the ultimate pointing of the bow.

AS: **And at one particular moment, when you most strongly feel that connection, you choose to release the arrow—or the shutter.**

TS: I found the idea very interesting: in order to make a photographic picture that speaks, you have to become the subject. You have to really love what you're looking at, and become one with it for that moment. You have to release yourself completely to that subject. I mean, it's a bit idealized—I don't want to dramatize it too much— but when I read that book, I thought: That's true. There's something there I identified with a lot.

AS: **Many people talk about that experience of feeling "one with" something in relation to nature. But do you feel that experience translates to when you're photographing a scientific laboratory or research facility as well, as you have been doing in recent years? When you're faced with man-made environments and technology, can you still get into that headspace?**

TS: Not really. My interest in technology was mainly driven by a feeling that there's a general, almost obsessive conviction about the advantages of technological progress, and an unwavering belief in the promises made by technology. I'm not against the development of technology in general, but society's blind conviction in terms of its benefits could potentially be very dangerous if sociopolitical development remains so far behind, as is happening right now. Technology is moving forward at the speed of light, but within the sociopolitical fields, and in terms of human coexistence, we are marching backward. I thought it would be interesting to make pictures that show this obsession with scientific progress— to look at the technologies and these elaborate scientific spaces as a representation of a mind-set. I'm trying to identify pictures that communicate their presence in the real world.

A shared phenomenon and principle in all fields of research and science is that people have to be extremely focused—they have to imagine and then explore unknown territories, squeezing them through a small pinhole in order to come up with a conclusion. It's like a riddle, and, in order to solve it, they build these crazy environments that cost billions of euros, and spend forty or fifty years working on it until they find a solution—if they ever find one.

That said, in doing so, they also have to come together and work at the same table—Japanese, Chinese, Germans, Ukrainians, Americans, Israelis, Romanians, Australians, Africans. That is fantastic, and we could ask ourselves why that's so much more difficult in political or humanitarian spheres. Look at the United Nations: Donald Trump stands before the U.N. and says every nation should think of themselves. What kind of signal is that, especially given the situation we're now in with respect to the environment? It's crazy.

AS: Given what you've seen, experienced, and learned while making *Nature & Politics*, is climate change of particular concern to you?

TS: Well, the question is: Can we save the globe or not? And, if so, what are we doing to save it? We all have the same problem and must act globally—it's an opportunity to be more united than ever before. Part of my desire to make work about science and technology was also to open doors and show the power of collective efforts. Maybe this is naive, but you have to have a reason to work.

AS: How do you initially explain your artistic intentions to the people who work within the scientific facilities that you photograph?

TS: I've encountered a huge amount of generosity, and what I've found is that, like artists, scientists and researchers are working toward the unknown. I would say I often found an akin mentality.

AS: In *Nature & Politics*, alongside the photographs you've made in research facilities and scientific laboratories, you also include pictures that you've taken of the original Disneyland in Anaheim, California. What is the relationship between all of these works?

TS: That thought process started with an article I read about Disneyland in a German newspaper about eight or nine years ago. At the time, I was thinking about how the movie industry today— with all its digital technology—can create realistic-looking footage of anything that one imagines. I started thinking about Disneyland looking like an archaic moment in the history of fantasy creation. Of course, since the beginning of film, there has been science fiction—*Metropolis* (1927) by Fritz Lang, and so on. The original Disneyland had something to do with imagination, memory turning into sculpture, and I thought it would be interesting to try to let today's technology setups and Disney associatively play with each other.

AS: When paired with the images of scientific research, these photographs force one to reevaluate some of your technology pictures through the lens of fantasy and science fiction. All of a sudden, things like *Solaris* (1972), or *2001: A Space Odyssey* (1968), or even the "Sorcerer's Apprentice" segment of Disney's *Fantasia* (1940) become reference points too.

TS: Yes, absolutely. But I never think about it as science fiction. I think more about it as the individual words—*science* and *fiction*— and how the two things are so closely intertwined. For example, in the photograph of the Matterhorn ride at Disneyland, the mountain is clearly not the real Matterhorn, so what does it show us? It's an embodiment of Disney's fascination with traveling in Europe, being overwhelmed by a natural phenomenon, and creating a papier-mâché masquerade of that experience for others. If you don't know exactly what the machines and environments within are doing, looking at laboratories and scientific research centers offers

I'm not against technology, but there is always a political agenda in place, which is what I question and try to make art about.

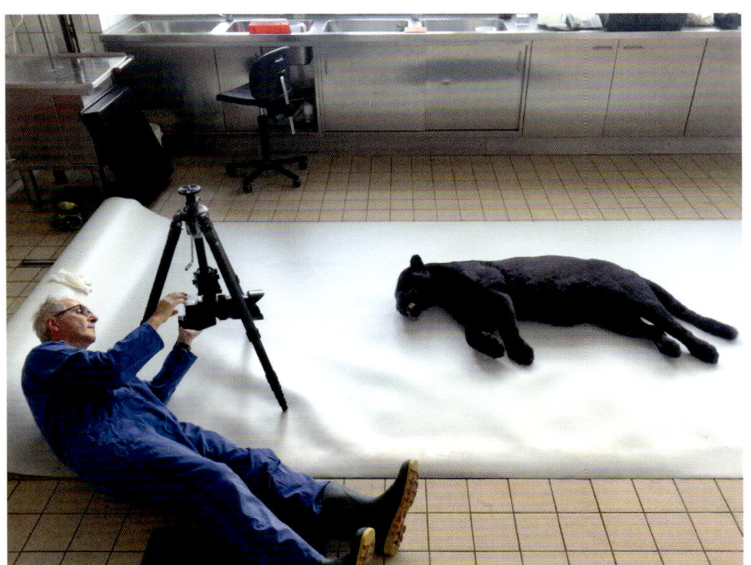

Thomas Struth making
Jaguar at IZW Berlin,
September 21, 2018
© Noah Mueller/Studio
Thomas Struth

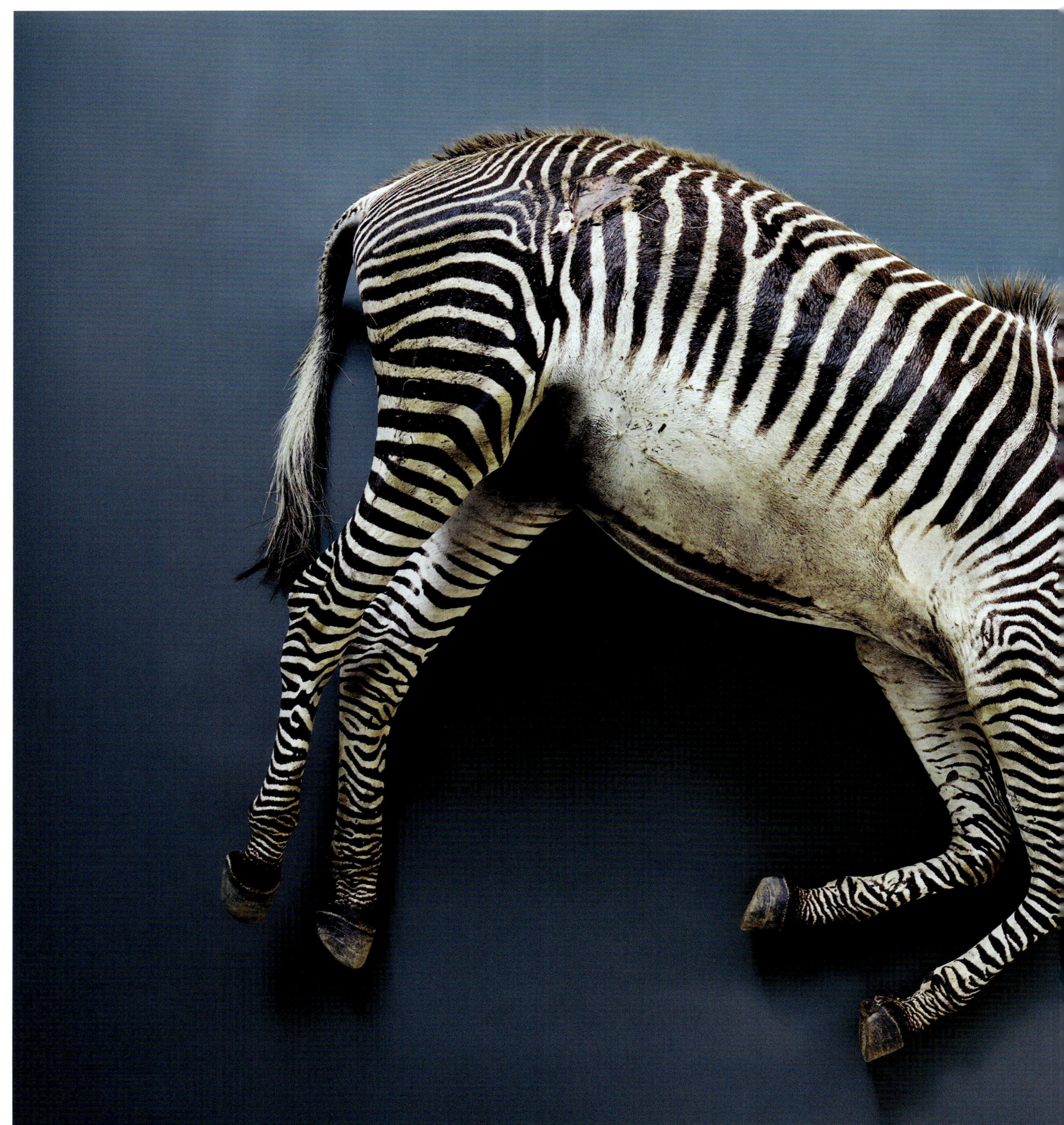

Zebra (Equus grevyi),
Leibniz IZW, Berlin, 2017
Unless otherwise noted,
all photographs © the
artist and courtesy Marian
Goodman Gallery, New York
and Paris

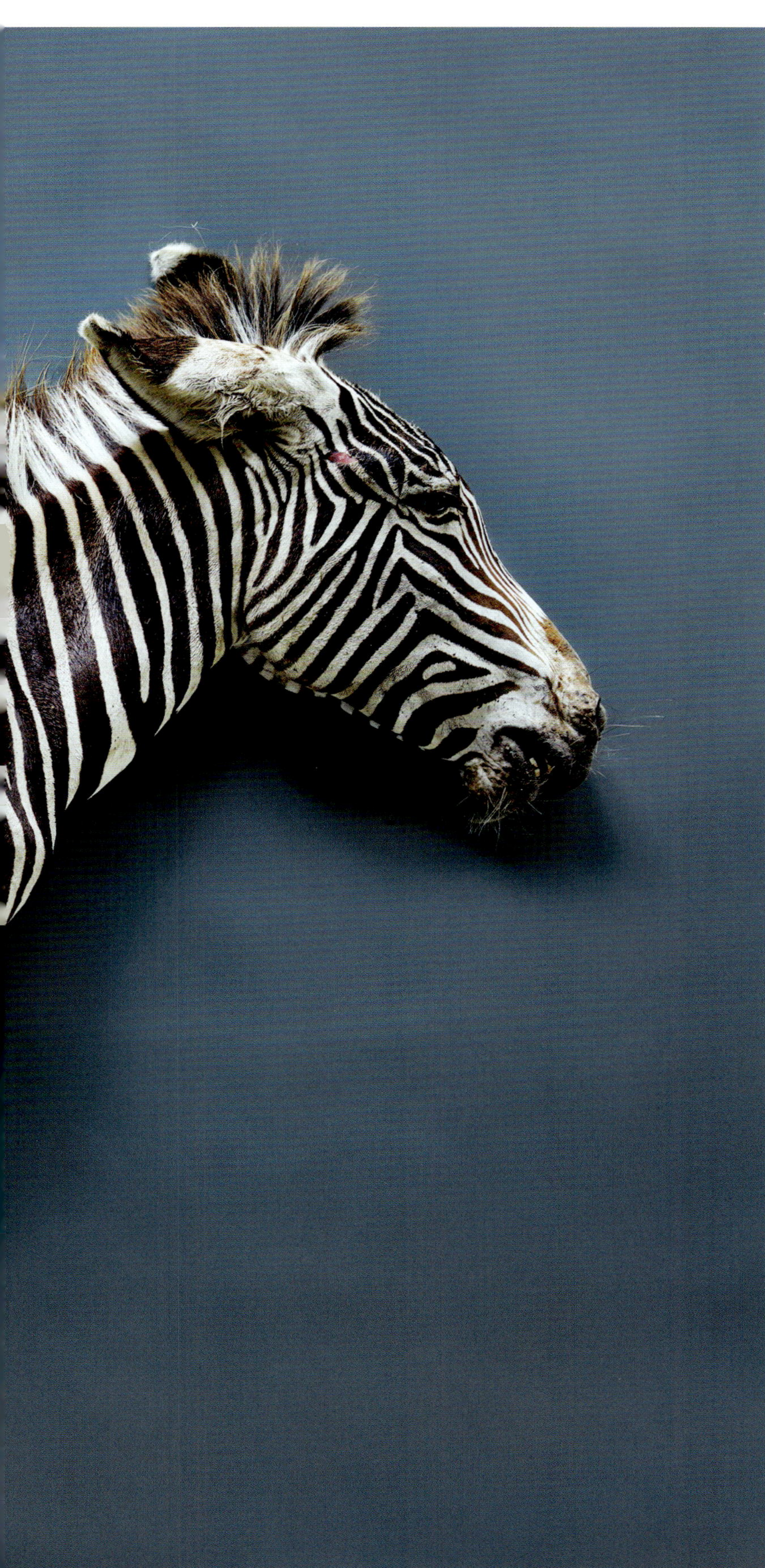

an opportunity to ask, What do they tell us as an atmospheric entity about humankind's aspirations, obsessions, entanglements? Again, I'm not against technology, but there is always a political agenda in place as well, which is what I question and try to make art about.

AS: **In a sense, your intentions and motivations are both political and personal.**

TS: Yes. In recent photography, the personal and private have become so dominant—through Instagram, social media, and photographers who have been very successful in celebrating their private lives. I find this a bit boring. In the history of art, everything that has survived from any culture was not art that remained within the private sphere.

Since the end of 2016, I've been photographing dead animals at the Leibniz Institute for Zoo and Wildlife Research in Berlin. I wanted to do this partly because I'm getting older, people around me are starting to die, and the limits of life have become more apparent to me than they used to be. The subject matter here is something that touches me and comes from my inner thoughts, but is also more generally about humankind at large.

AS: **Looking back at your overall oeuvre, this new body of work that you're making at the Leibniz Institute seems to take a very different approach, at least aesthetically.**

TS: Yes. It's very different, and I'm surprised myself. That said, I have made work within the medical field in recent years—in relation to technology's response to sickness and disease—so I had already come close to death as a subject matter. I never felt like photographing dead people, however, because this would always be a specific individual. When it's a dead bear or zebra, it's more of an "animal" or a "soul."

AS: **Do you see these photographs as being along the lines of memento mori?**

TS: The work is definitely in line with the history of memento mori. Maybe twenty years ago, I was in Milan at the Pinacoteca di Brera and was quite struck by Andrea Mantegna's fabulous painting *Lamentation over the Dead Christ* (ca. 1483). The body has this gesture of a complete absence of tension, which kept coming back to my mind. I'm still in the process of slowly making this work.

AS: **Last week you were photographing the corpse of a black jaguar at the Leibniz Institute. Do you feel that this particular picture was successful?**

TS: I think so, but I haven't edited the results yet. I made one picture from directly above, but then tried something else. Like in the Mantegna painting, I went very low and photographed the jaguar from a very short perspective—I wanted to concentrate on its face—and, in the back of the picture, you can see the legs of the lab tables, as well as the feet of some of the people who were standing there. The gesture of the animal's body means a lot. With only a little change to the camera position, the body suddenly looks like it's jumping or falling. Very minute reorchestrations of perspective completely change how the animal appears in the frame. It's very peculiar to be in the presence of death.

Aaron Schuman is a photographer, writer, lecturer, and curator based in the U.K. His latest book, *SLANT*, will be published by MACK in 2019.

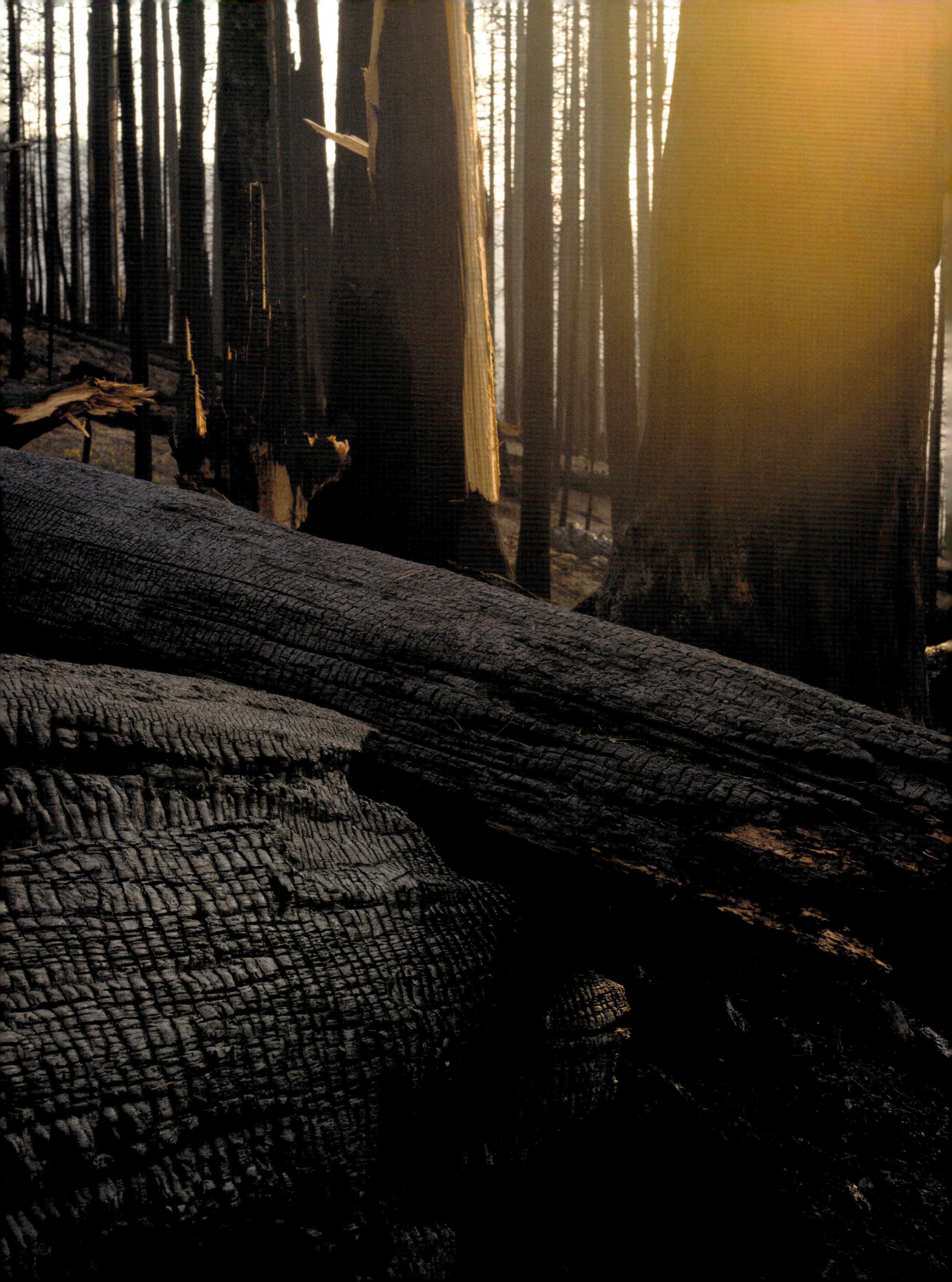

Carolyn Drake

William Finnegan

California Burning

Every fire has a narrative. For major wildfires, there are numbers to help frame the narrative, and usually a name. But all fires start long before they start, in the sense that the ground must be prepared, literally, for the conflagration to come. Fuel, weather (and behind weather, climate), the natural landscape, the built landscape, suppression efforts past and present, prevention schemes, politics—these factors and many others, interacting and colliding, create a context and prehistory for each major fire. The fire itself burns and then takes its place both in recorded history and in the natural history of its epoch. That epoch, now, is the Anthropocene—the epoch of a world made by humans.

To make these images in California's disaster-struck areas, Carolyn Drake chose the long moments after a series of major wildfires. The flames and smoke, the panic and news crews, are gone. The land is charred, ashes are sifted, burned-out residents return, campgrounds reopen. The fury and violence of the vast event recede, and the world that's left behind becomes specific again, inviting contemplation. Three big wildfires broke out in Northern and central California in July 2018 and burned through the following weeks. Drake tracked and photographed the aftermaths of all three.

The Ferguson Fire was started by a vehicle's overheated catalytic converter in dry vegetation beside a highway in Mariposa County. This was in the Sierra foothills southwest of Yosemite National Park. The fire burned into the park, and its smoke filled Yosemite Valley, the most popular (and most photographed) destination in the Sierra Nevada mountains. The valley was closed and evacuated. It became a staging area for firefighting operations. Roughly three thousand firefighters were thrown at the fire. Two died. One was a bulldozer operator, killed when his vehicle rolled down a mountainside. The other, a captain of an elite crew whose members work in the most dangerous areas of wildfires, was killed by a falling tree. Nineteen other firefighters were injured. The Ferguson Fire took more than a month to contain. It burned almost ninety-seven thousand acres.

The Carr Fire was started by a flat tire, which sent a wheel rim onto asphalt, where it generated sparks, igniting dry vegetation. This occurred in the mountains west of the city of Redding. Three days later, the fire jumped the Sacramento River and entered Redding, forcing the evacuation of thirty-eight thousand people. That same evening, a fire whirl developed— a tornado-like column of superheated air that can be generated by intense wildfires. The Redding fire whirl contained winds exceeding 143 miles per hour. The wind tore roofs from houses, bark from trees, and toppled high-tension power-line towers. The fire whirl was reported to be forty thousand feet tall. The Carr Fire killed eight people, including three firefighters. It burned more than a thousand homes. Insured losses were estimated at $1.5 billion. The cost of fire insurance is said to be soaring in California, moving beyond the reach, effectively, of many residents in fire-prone areas.

Finally, the Mendocino Complex Fire started as two vegetation fires in the chaparral-covered mountains near Clear Lake, about one hundred miles north of San Francisco and fifty miles from the coast. The ignition point is still under investigation. In hot, dry, windy conditions, the fire burned for nearly two months, ultimately consuming more than 450,000 acres, which makes it the largest wildfire in California history. It destroyed 280 structures, most of them rural residences, many in unincorporated communities like Spring Valley. One firefighter died. He was a battalion chief from Utah, killed by a falling tree. The firefighters struggling to contain the Mendocino Complex Fire were hampered by a manpower shortage—many of their brethren were off working other fires. Nevada state prison inmates, among others, were deployed to fight the enormous blaze, and suppression costs ran to more than $200 million.

Wildfires are getting bigger, hotter, more frequent, more destructive. Of the ten most destructive fires (measured by destroyed structures) in the history of California, six have occurred in 2017 or 2018. The single deadliest and most destructive, the Camp Fire, in Butte County, is still smoldering as I write, in November 2018. This fire, which essentially destroyed the small city of Paradise, California, on November 8, killed eighty-eight people, with 249 more still listed as missing. It burned more than thirteen thousand homes, and archaeologists are working among the ruins now to find any traces, such as teeth, of the perished. The second most destructive California wildfire, the Tubbs Fire, in Napa and Sonoma Counties, did its gruesome thing in October 2017. It killed twenty-two people and burned more than five thousand structures, including some 2,900 homes in the city of Santa Rosa.

Why this biblical plague of fire? It is the Anthropocene, and we must look to our own agency. The climate is hotter and, in California, much drier than in the recent past because of the greenhouse effect, which is caused, primarily, by the burning of fossil fuels. In the American West, as in other places, logging practices have produced vast amounts of slash—woody debris that burns more readily than the mature forests it replaced. Overzealous fire suppression policies have contributed, paradoxically, to the proliferation of bad fires. They have left a tree-choked landscape, where natural fires caused by lightning have been unable to do their ecological job of thinning.

Then there is the mass migration of people into what land-use jargon calls the wildland-urban interface (WUI)—the zone where residences abut forests or other combustible vegetation. This is by far the fastest-growing land-use type in the United States. More than one hundred million souls now live in the American WUI. These people accidentally start a great many fires, and their presence makes firefighting harder and more dangerous. Natural fires cannot be allowed to burn themselves out anywhere in or near the WUI.

These calamitous trends and policies form some of the context, the prehistory, of the wildfires that rage in our day. These problems are national, if not international; climate change is, of course, global. But here is Stephen Pyne, the preeminent American scholar of fire, on California, its particular problems, and the larger picture:

California is a special case. It's a place that nature built to burn, often explosively. If people vanished, fires would still thrive…. But people have worsened the scene. They have introduced flammable grasses, overgrazed in the mountains and felled forests in ways that overturned the prior system of ecological checks and balances…. And then Earth's keystone species for fire decided to burn fossil biomass, which has cascaded effects throughout the planet and unhinged the climate. We used to think fire history was a subset of climate history; now climate history is becoming a subset of fire history.

I see this dilemma, this mess, in Drake's pictures. They are not about simply the harsh aftermath of natural disasters. They are about the role of humans, and of Drake's own observing eye, in this burning world we've made.

William Finnegan is a staff writer at *The New Yorker* and winner of a 2016 Pulitzer Prize for his memoir *Barbarian Days: A Surfing Life*.

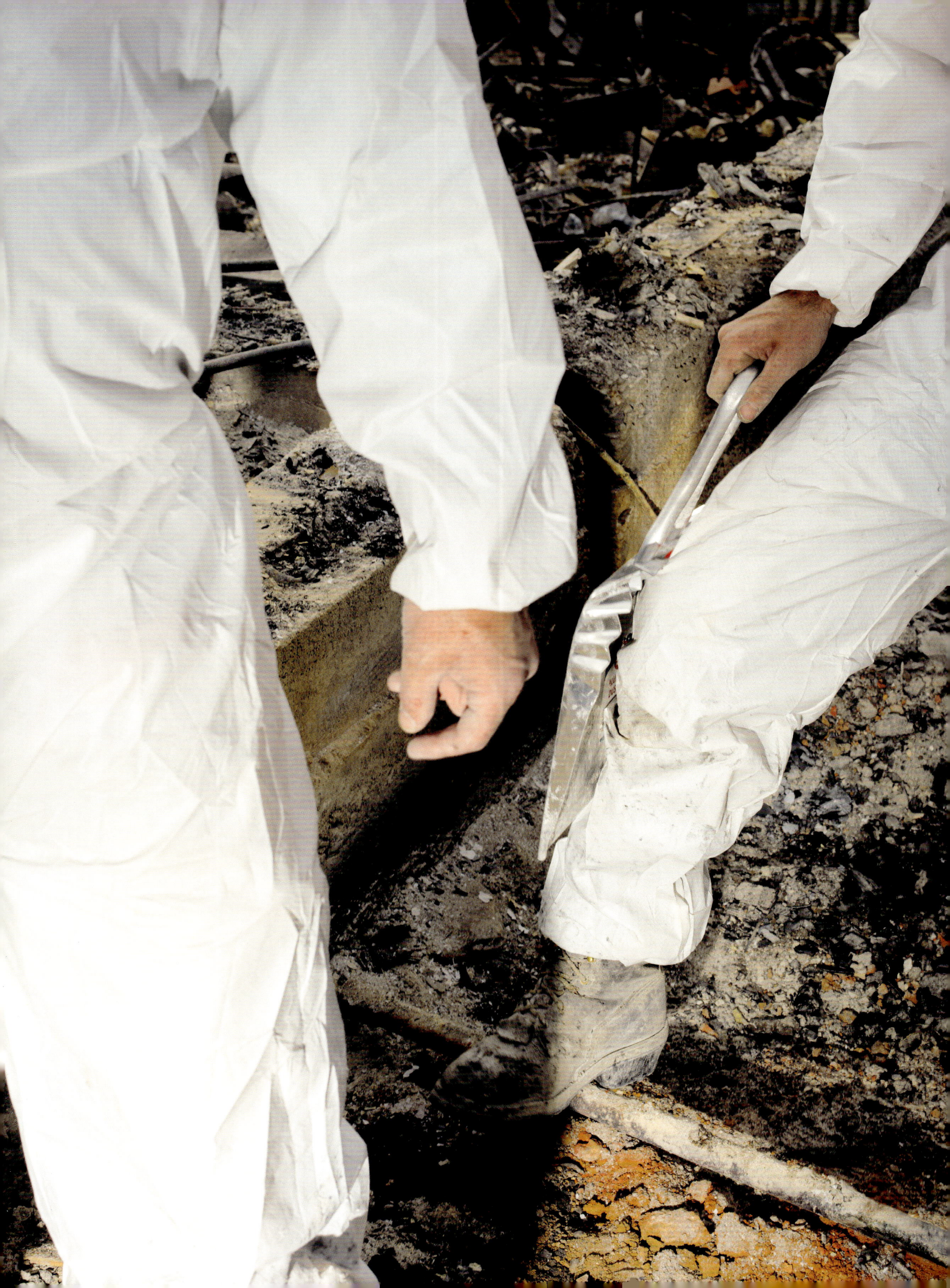

Opposite:
Archaeologists search
for a jar containing the
ashes of a widow's husband
after her house burned
down, Redding, California

This page:
Fire roads cross privately
owned mountains, three
weeks after the Mendocino
Complex Fire finished
burning, Spring Valley,
California

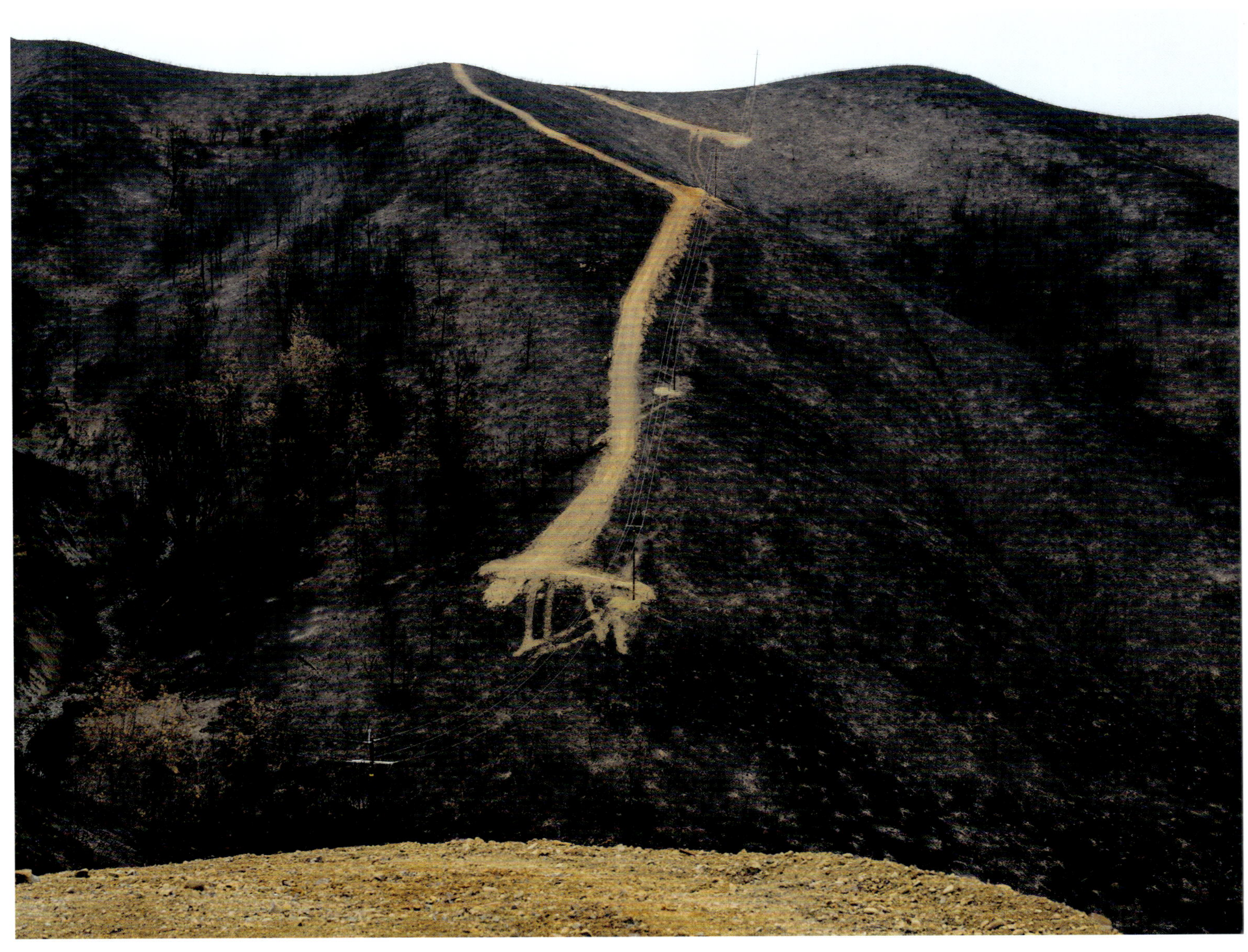

Lounge in the Usona Forest
Fire Station, Mariposa
County, California

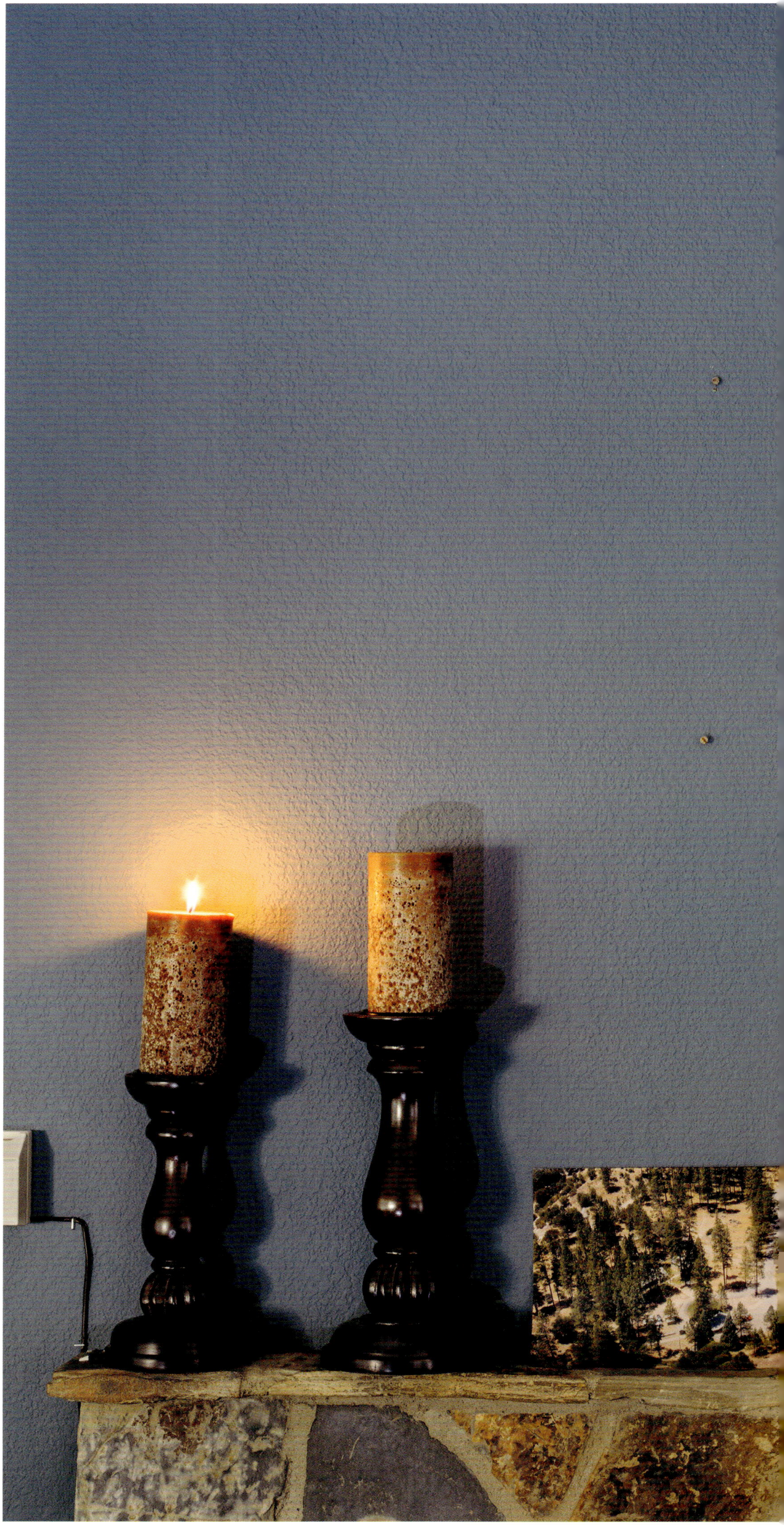

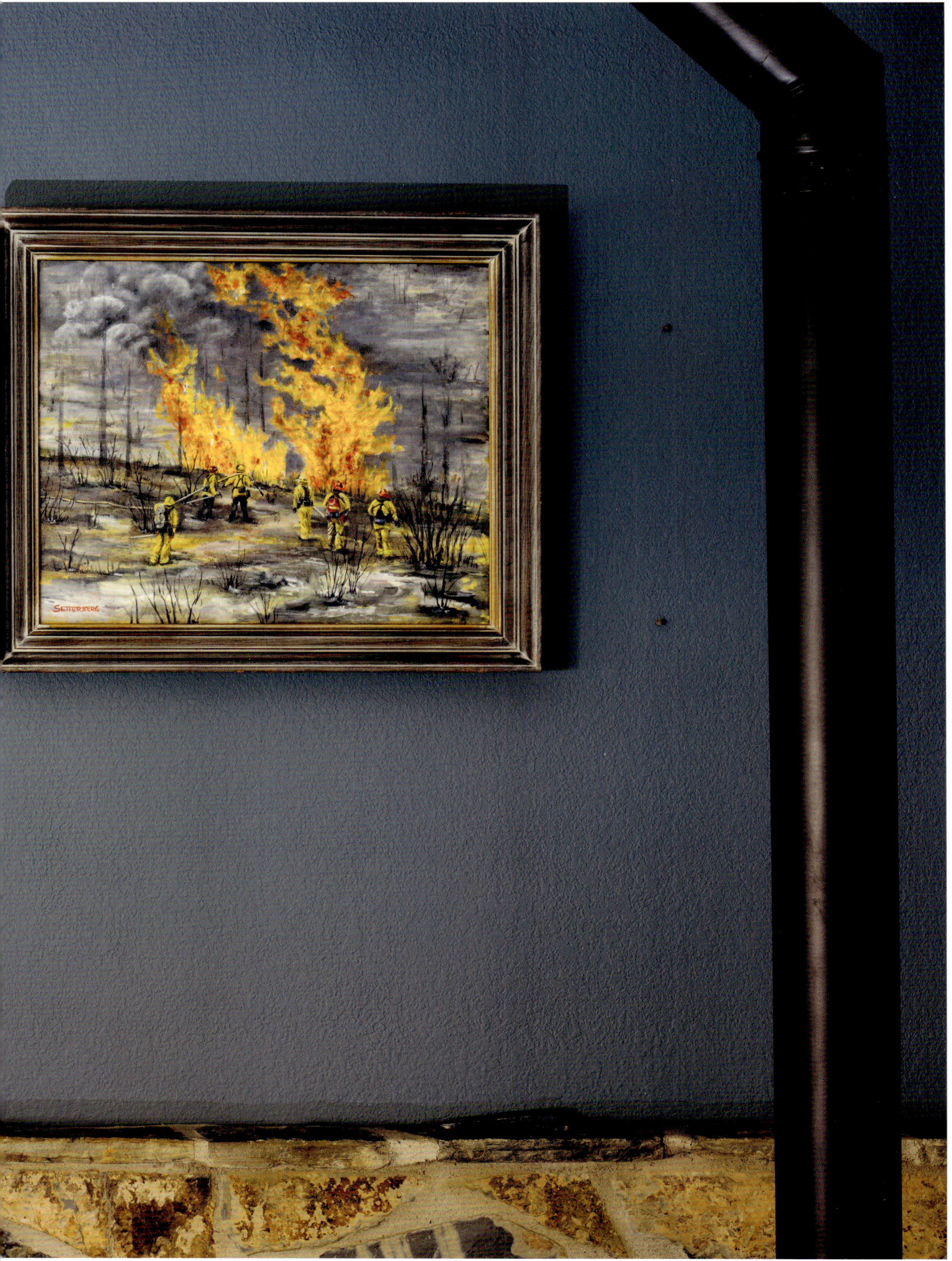

This page:
Rod on the foundation of his burned home after the Mendocino Complex Fire, Spring Valley, California

Opposite:
Creekside shrub, one month after the Carr Fire, Redding, California

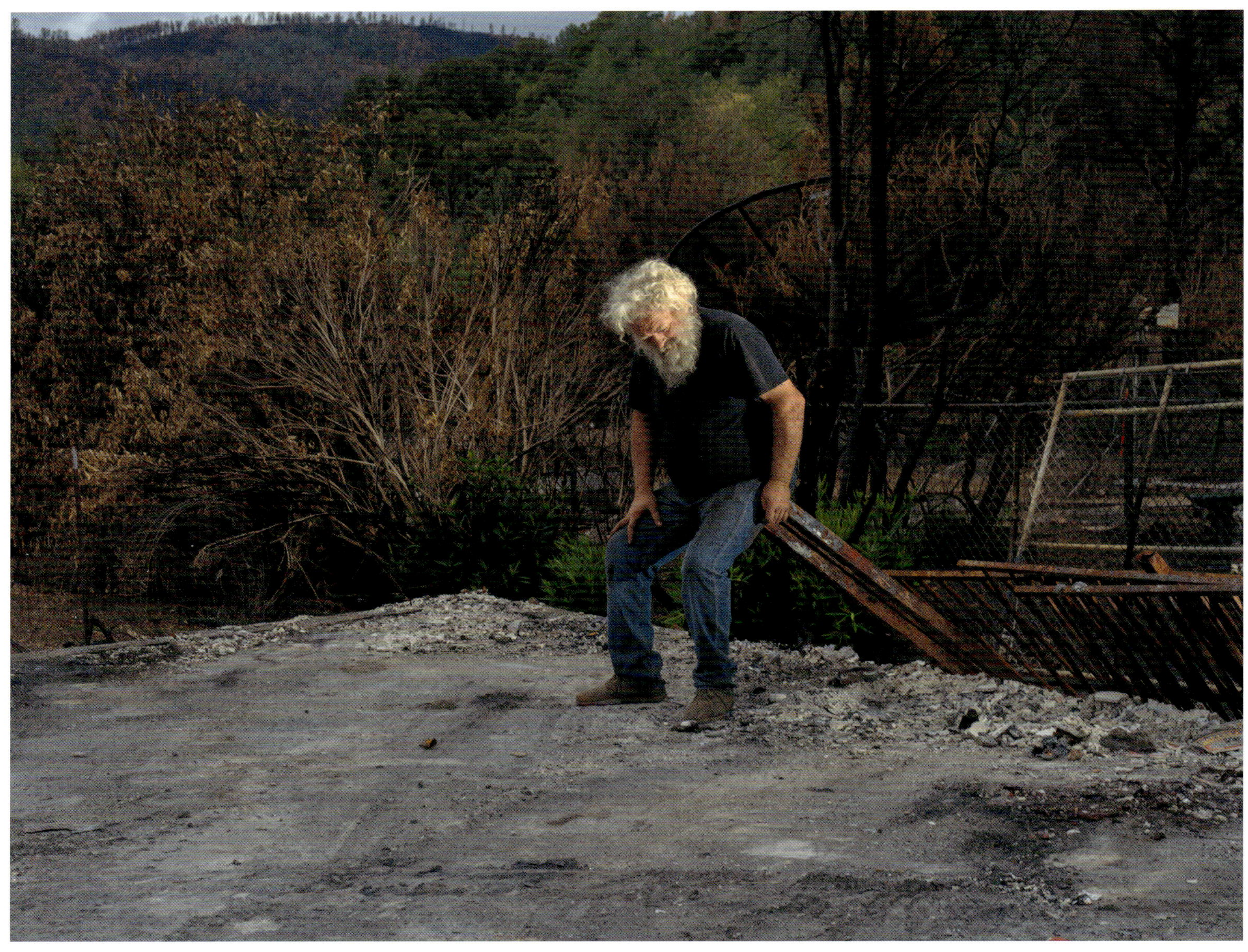

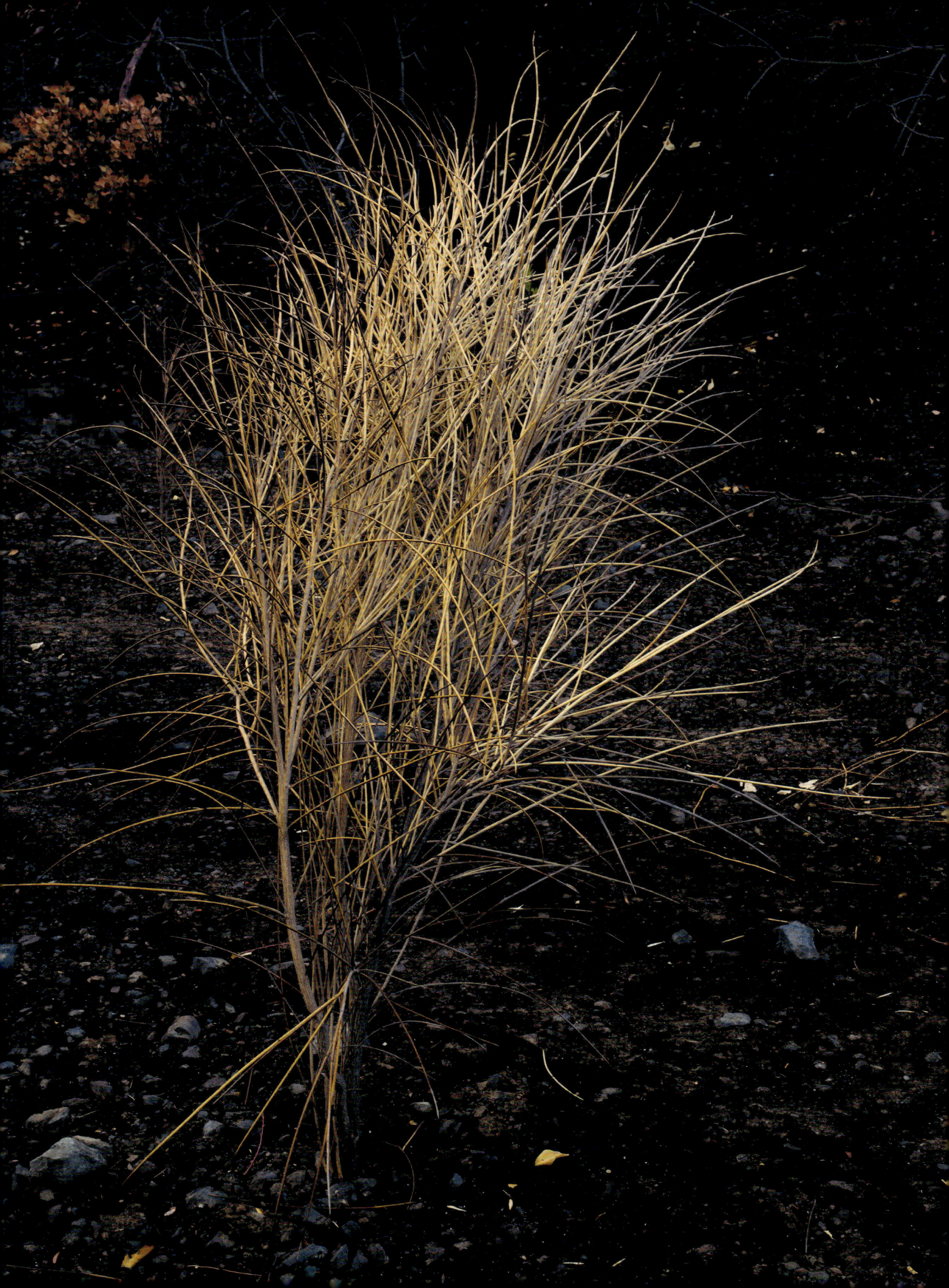

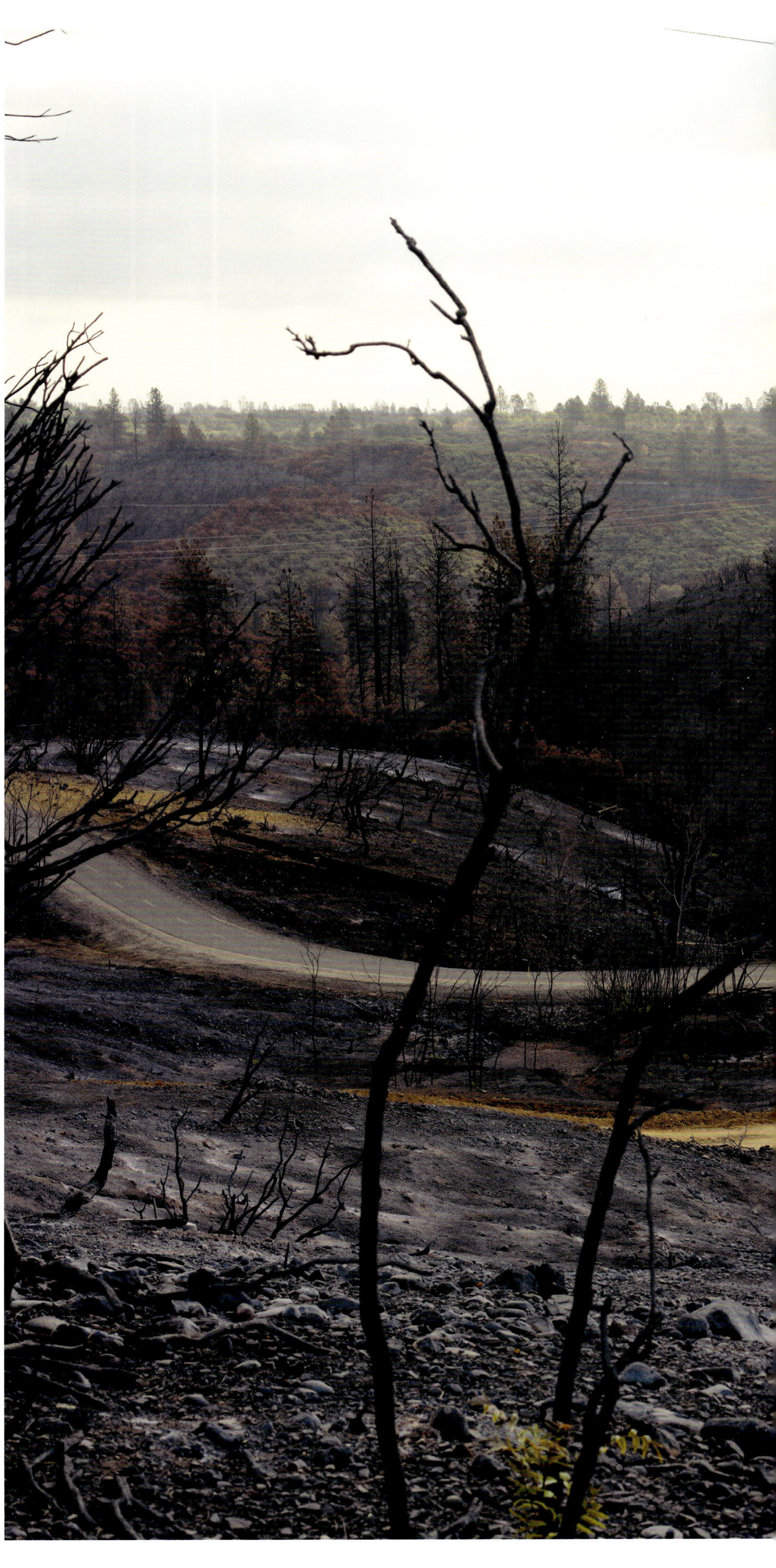

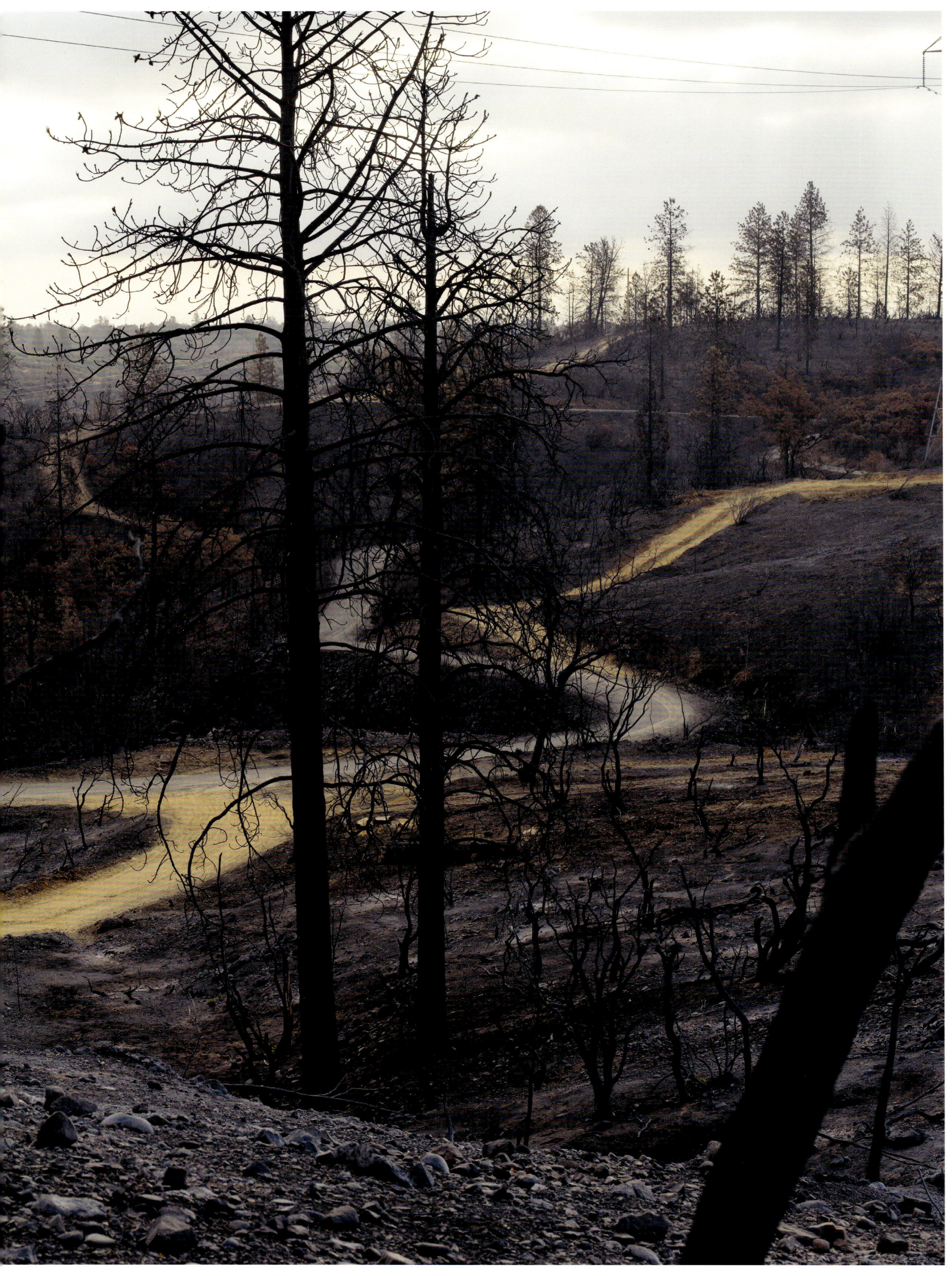

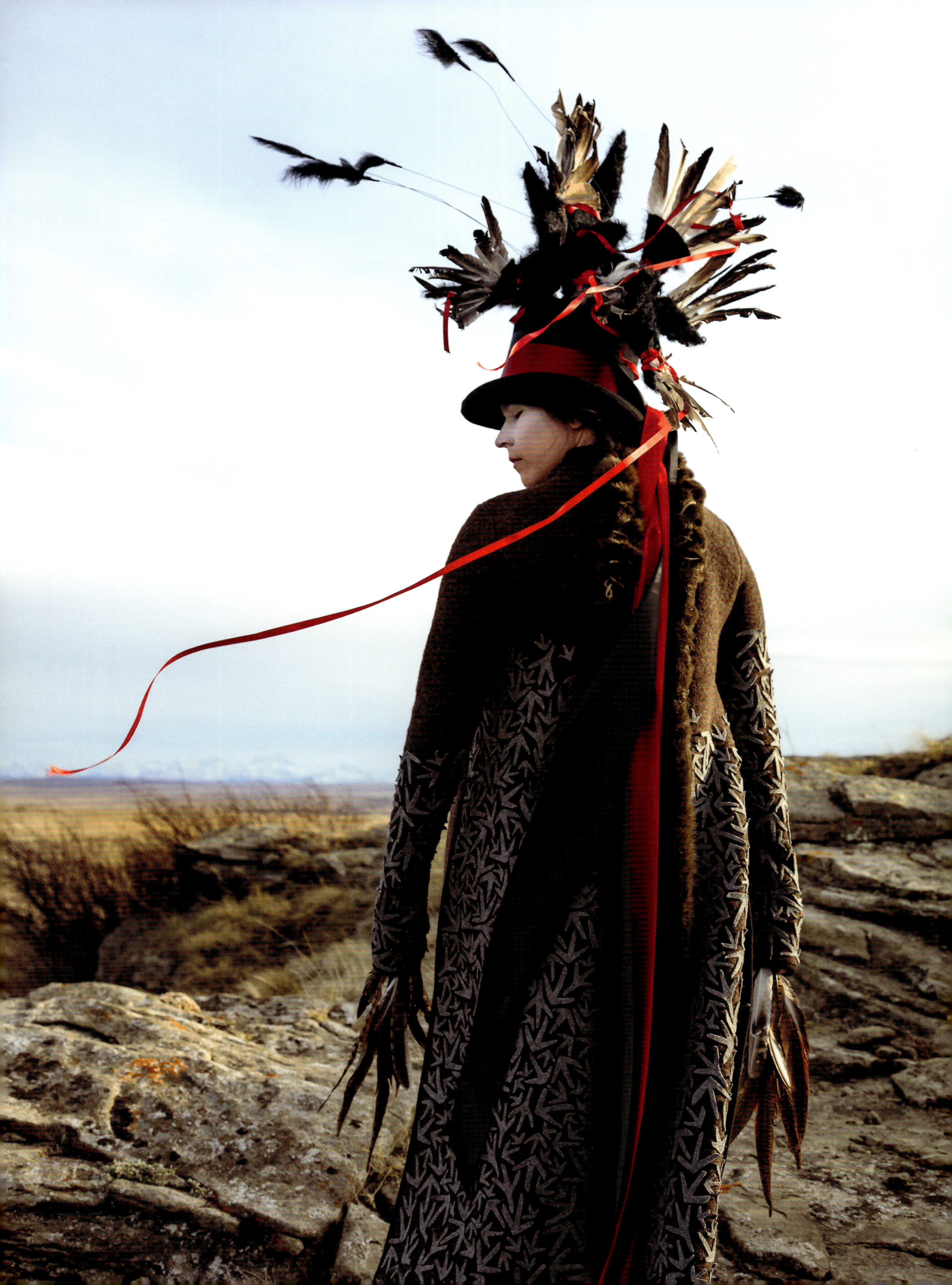

Wanda Nanibush

Notions of Land

For Indigenous artists, how can photographs provide
a space of visual sovereignty?

**Meryl McMaster, *Bring me
to this place*, 2017**
Courtesy the artist; Stephen
Bulger Gallery, Toronto; and
Pierre-François Ouellette
art contemporain, Montreal

Imagine a lonely warrior slumped over his horse, backlit by a sun setting over rolling hills, or a group of warriors riding off toward the horizon. It isn't hard for most people to conjure because we have been raised on these romantic images of the so-called vanishing Indian. Photography, which developed hand in hand with colonialism, has largely been responsible for the continued stereotype of the noble savage. What "Indians" are admired for—the idea of being one with nature, one with the land and animals—is also seen as the source of their inferiority and inevitable demise. It is given as the main reason they are unable to survive in modern society: "Indians" are part of nature, not civilization, and, by extension of this argument, less than human.

American photographers in the nineteenth and early twentieth centuries, like Edward S. Curtis and Joseph Kossuth Dixon, to name only the best-known, used photographs to link Indigenous Peoples to the idea of "nature" in order to speak about the end of Indigenous "nobility"—an end that was connected to the introduction of supposed "civilization" in the form of white-settler communities. Photography itself was considered part of the proof of white superiority because of its basis in technological innovation.

All stereotypes have a minor truth to them: in this case, it's true that Indigenous Peoples have a kinship with land and animals. We do think that all living things have a spirit. The idea that the earth is our first mother and that animals are our kin and can communicate with us through visions, dreams, and signs is central to many Indigenous cultures. But is this all based on clichés and superstitions? No. These facts were deeply misunderstood and parodied for the sole purpose of justifying the removal of bodies blocking the path of settler colonialism; today, this view is used to rationalize the extraction economy as well. In recent photographic works, a number of Indigenous artists are beginning to debunk these colonial "facts" and counter them with deeper understandings of our relations with "nature."

Robert Kautuk, from Kangiqtugaapik (Clyde River) on Baffin Island, in Nunavut, Canada, uses photography to document the daily life of an Inuit community, in particular their relationship to the land, weather, and animals. For me, a southerner, viewing his aerial photograph *Walrus hunt near Igloolik, Nunavut, Canada* (2016), of an ice floe covered in blood after a successful walrus hunt, is almost surreal. South of the Arctic, this is not what our hunting looks like. In Kautuk's photograph, the red bloodstain spreading across beautiful white ice is set against the deep blues of the Arctic Ocean. There are minute details of hunting gear, a small boat filled with necessary supplies, and people braiding intestines. You want to look closer at this almost dollhouse-scale perspective, which was taken with Kautuk's Phantom 4 drone rather than his digital single-lens reflex camera.

Kautuk also employs his skills on a digital mapping project called the Clyde River Knowledge Atlas (2015–ongoing), which draws on elders' and harvesters' knowledge of local Inuit place names and the environment. The atlas is an important tool for sharing resources on environmental assessments, changes, and uses, and the names for places in the Inuktitut language are critical for determining the locations of people who find themselves in an emergency. Kautuk's photographs and the new atlas may also have a lasting impact on Inuit communities' ability to assess the desirability of outside development projects. In 2017, Clyde River won a Supreme Court of Canada fight over the National Energy Board having granted permission to energy companies to do seismic testing for underwater oil in Nunavut. The court justices ruled that the NEB had trampled Clyde River residents' rights as Indigenous Peoples and had failed to consider the necessity of subsistence hunting for Inuit existence. Clyde River's former mayor Jerry Natanine, who fought for years to have this Inuit

way of life acknowledged and respected, felt vindicated by the ruling.

Kautuk, in his way, is doing the same by preserving the beauty of the land and of the hunt; *Walrus hunt near Igloolik, Nunavut, Canada* gives intense presence to Inuit survival. Unfortunately, the hunt, especially of seal, has long been misunderstood by southerners and Europeans, often in the name of environmentalism and ecology. Many people do not understand that grocery-store food is very expensive in Nunavut and often just not available, so hunting is the only way to eat. A diet of whale, walrus, seal, arctic char, and bear is also healthier physically and spiritually for the Inuit people. Hunting is tied to cultural practices that keep alive ways of being that hone the collective spirit through sharing, honoring, singing, and remembering the hunt. In this context, a photograph of a walrus hunt is an act of resistance.

The ways in which photography can express Indigenous presence on the land also include more conceptual, performative work, such as Shelley Niro's series *The Shirt* (2003). Niro is a member of the Turtle Clan of the Kanien'kehá:ka (Mohawk) Nation. *The Shirt* is a unique set of nine photographs that, taken together, create a narrative of Indigenous sovereignty where women are central. Niro often constructs her photographs by having people perform for the camera; in this case, her friend and fellow photographer Hulleah J. Tsinhnahjinnie, of the Taskigi Nation and Diné Nation, faces the lens, confronting the viewer directly. She is photographed in the landscape wearing a series of five T-shirts that sequentially say: "The Shirt"; "My ancestors were annihilated exterminated murdered and massacred"; "They were lied to cheated tricked and deceived"; "Attempts were made to assimilate colonize enslave and displace them"; "And all's I get is this shirt." In the sixth image, she appears without any shirt; in the seventh, a smiling white woman wears the final shirt of the series. These seven photographs are flanked by images of the land, underscoring the importance of land rights to Indigenous struggles for self-determination. In Niro's work, the shirt becomes a souvenir from the highway of colonialism, ripped off the backs of Indigenous women who live there. Because the shirt is an object many people have encountered and is a form of self-expression, statements about the colonial land grab become readable and relatable to a broad, diverse audience.

According to Niro, "*The Shirt* series came about as I flew over the Texas landscape. I looked out of my window and saw the land below chopped up into squares, each square neatly fenced off from the other. I thought about the 'Indians' who fought for that land, as well as the sacrifices made by tribes and nations in their efforts to keep away the settlers from their land and communities. Hulleah's presence gives the series seriousness and strength." The presence of a Diné woman in this terrain also draws attention to the connection between violence against Indigenous women and the land: in an extraction-based economy, "man camps"—sites of temporary worker housing—spring up, and violence against the local Indigenous women also rises. Historically, as well, colonization has specifically targeted women, reducing them to the property of men under many policies and laws, including the 1876 Indian Act in Canada. Niro "believes narrative delivers a personal contact with the viewer"; her work asks viewers to be actively and empathically engaged, to place themselves in relation to the narrative as perpetrators or survivors of colonialism.

Walking the line between construction and documentary is Onondaga photographer Jeff Thomas, who grew up in Buffalo, New York, and is a member of Six Nations of the Grand River. When Thomas visited the Six Nations reserve in Ontario, Canada, in the 1970s, he was taken with photographing the homes and

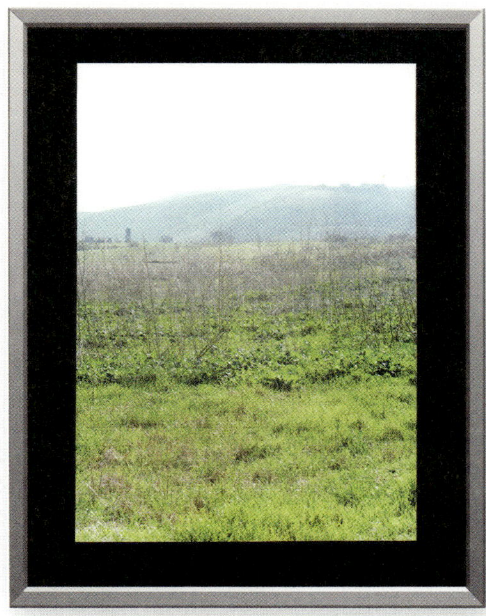
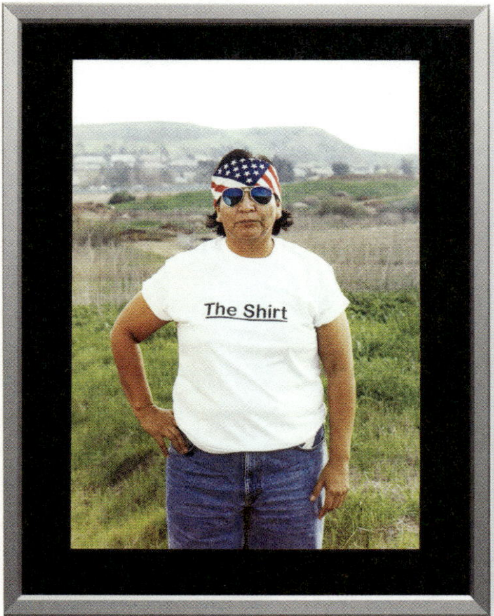

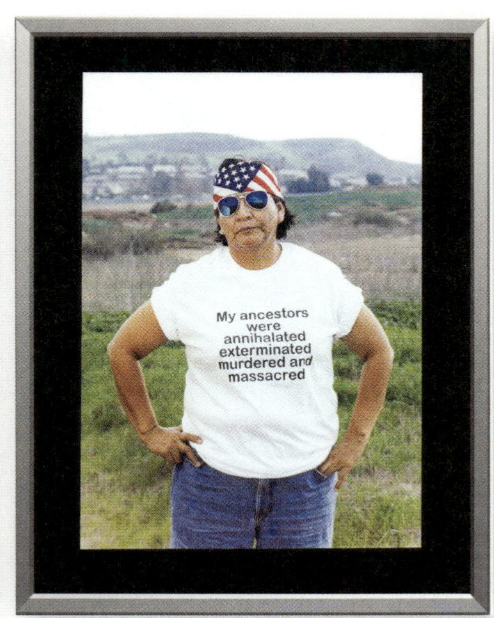

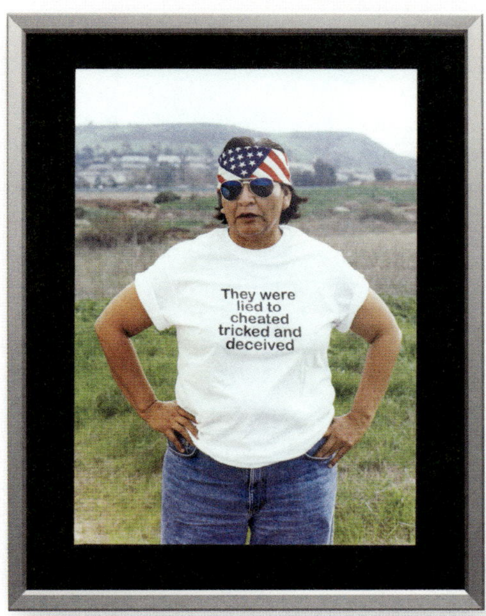

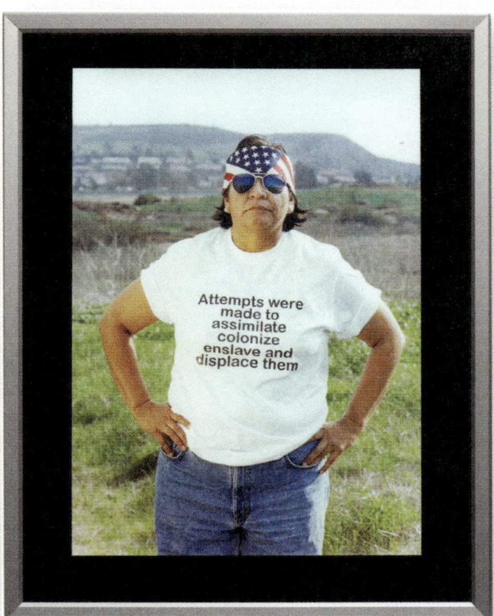

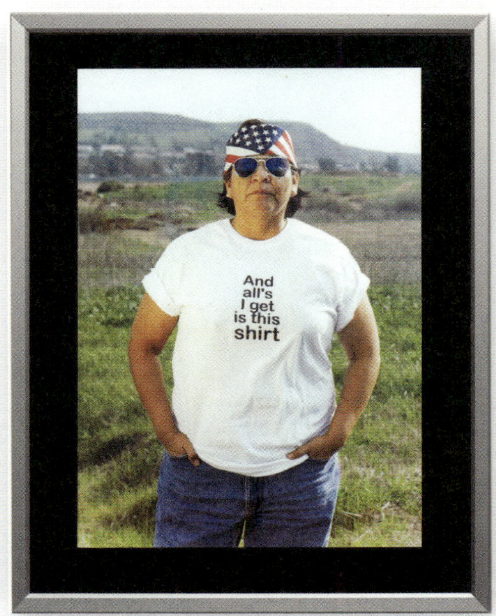

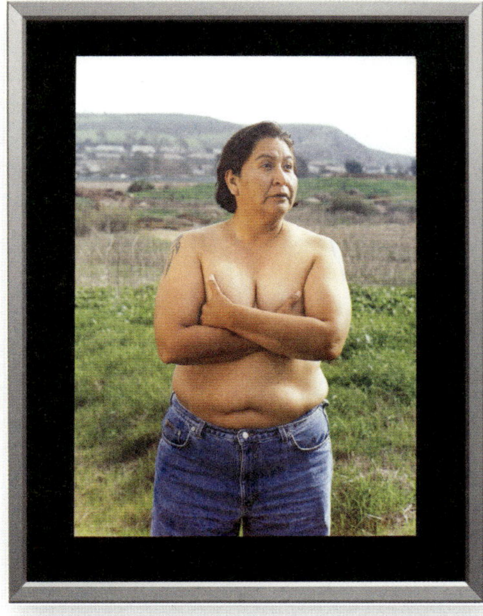
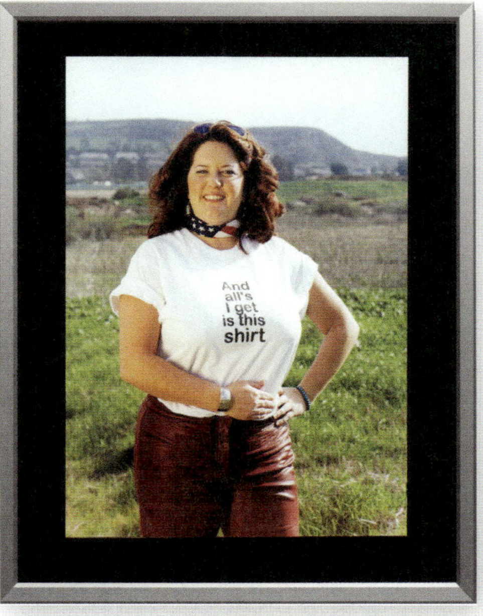

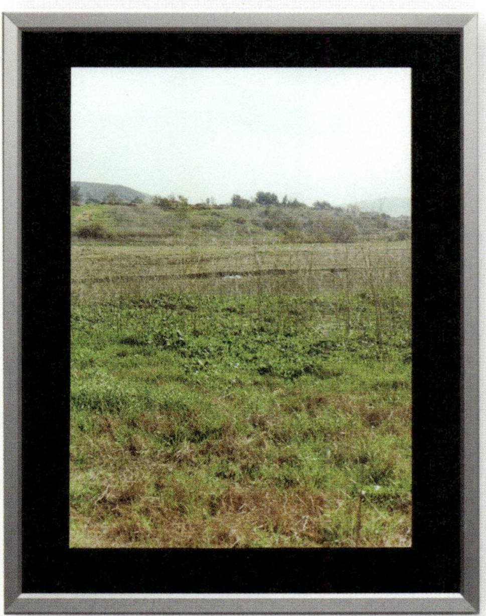

Shelley Niro, *The Shirt*, 2003
© the art st and courtesy Art Gallery of Ontario

Photography in the hands of Indigenous artists forms a body of philosophical, poetic, and physical knowledge of our relationship to land.

daily lives of elders, resulting in his series *Corn Husks* (1976–2011). He states that his step-grandfather "demonstrated how to weave the corn leaves into long strands…. The act of weaving corn leaves into a braid is a symbol of the Indigenous teaching that all living things are interconnected." Thomas purposely photographs the process of creation rather than the finished product, citing the bodily memory that is the basis of cultural knowledge. In the image *Bert General Husking White Corn, Six Nations of the Grand River* (1980), the movement of ripping husks from corn is culture in action. Thomas sometimes juxtaposes his images with archival photographs to create more complex narratives; *White Corn* (2014/2017) is composed of three photographs by Thomas and a fourth taken in 1912 by English anthropologist Francis Knowles, titled *Chief Jacob General*. These photographs are laid out like a wampum belt, a traditional medium for passing on oral history; Thomas uses photography as memory in place of wampum beads.

The way the physical environment can evoke centuries-old knowledge and function as a kind of memory making for Indigenous Peoples also informs the photographic practice of Kanien'kehá:ka (Mohawk) artist Greg Staats. In the striking work *untitled (restraint_ constraint)* (2015), he delves into the psychological aspects of place. A tree trunk with roots still embedded in the earth but with many severed parts fills almost the entire image plane; only a slight vision of concrete appears in the background. The work evokes a liminal space between the grounded and the uprooted while giving no particular markers of place. Staats describes his practice as "in service to orality of place and natural-world mnemonics—objects/images and place share a long wordless relationship—reunited to further their cause of holding together

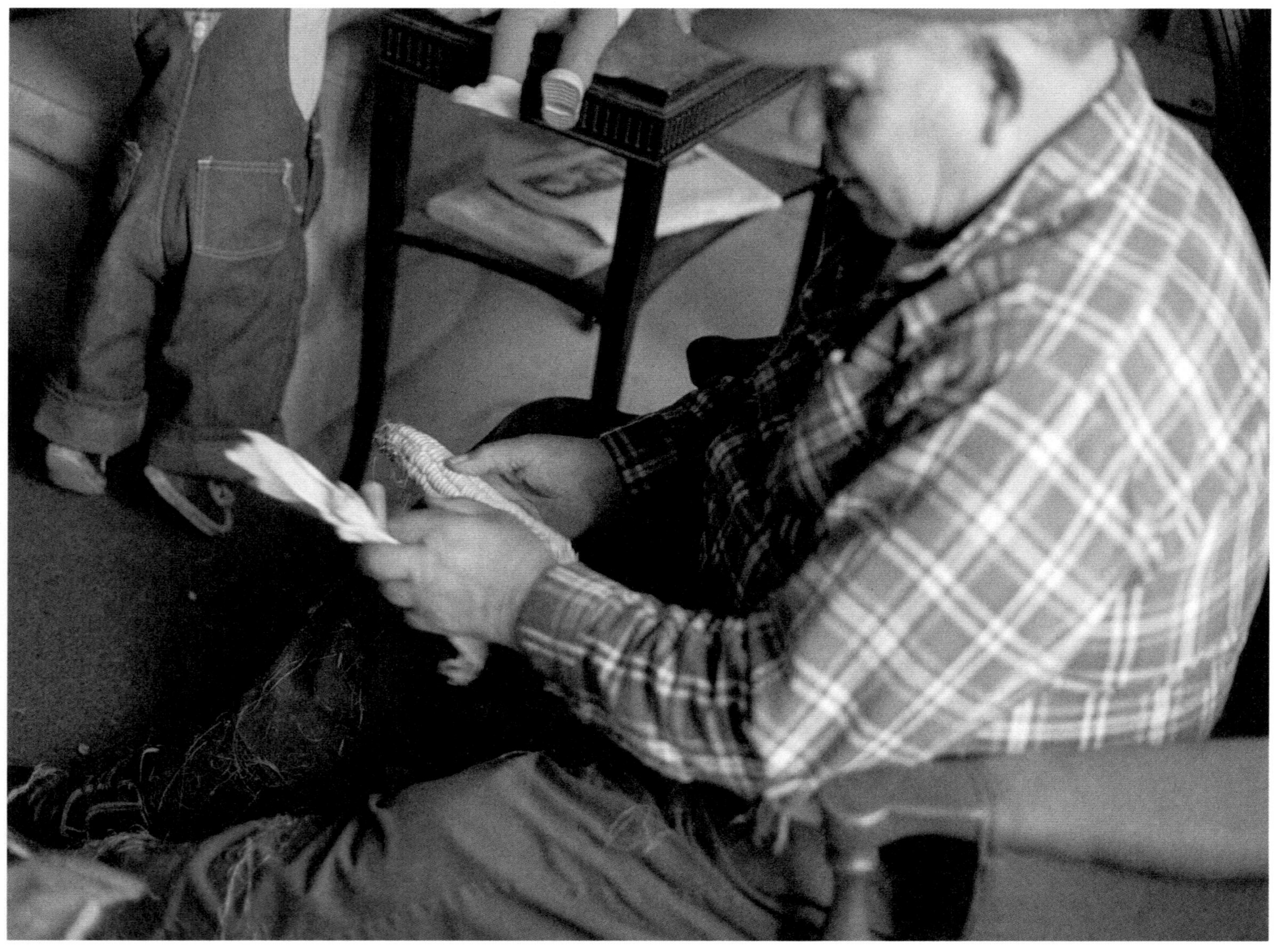

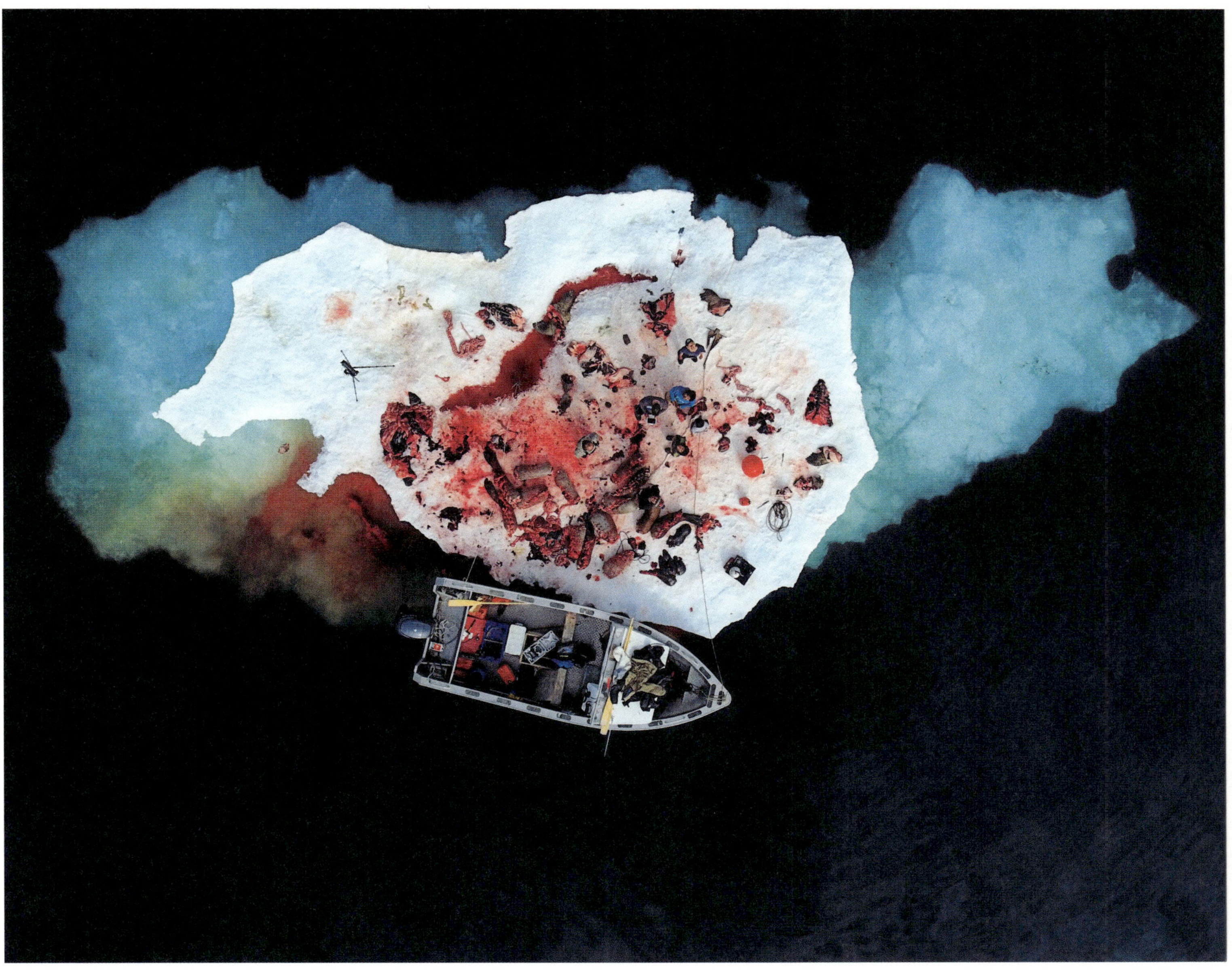

the unsaid." The trunk becomes a poetic rendering of Indigenous Peoples finding their way back home through the trauma of colonialism, which has left many with severed roots.

The land as a carrier of traumatic memory also forms a backdrop for Cree artist Meryl McMaster's latest photographic series, *Edge of a Moment* (2017), made in Alberta at the Head-Smashed-In Buffalo Jump. This spot was once key to Indigenous survival, especially for the Cree and Blackfoot of the prairies. Here, they would run buffalo off the cliff in huge herd hunts; such hunts occurred up until the 1880s. Settlers and the Royal Canadian Mounted Police massacred the buffalo almost to extinction to starve Indigenous Peoples onto reserves. McMaster uses herself as a model, constructing elaborate costumes to wear in these conceptual photographs. In the series, she a wears a coat covered in a printed chicken-feet pattern along with a hat fabricated from prairie chicken feathers. McMaster explains, "The feather bustle we see male traditional dancers wear is modeled after the chicken's tail feathers. I use their tracks in an abstract way to cover the garment I am wearing within the image. Their absence represents to me not only the dangers of the unsustainable use of the land, but also the human consequences of colonization and settlement." As McMaster makes clear, photography in the hands of Indigenous artists forms a body of philosophical, poetic, and physical knowledge of our relationship to land and the history of colonial ruptures. Photographs provide forms of visual sovereignty and assert a continued presence on the land, despite centuries of theft and removal.

Wanda Nanibush is curator of Indigenous art at the Art Gallery of Ontario.

Jochen Lempert

Brian Sholis

This portfolio by German artist Jochen Lempert includes a photograph, *Untitled (Botticelli IV)* (2018), of a detail from a Botticelli painting. It also includes *Anna Atkins* (2011), a photogram of a computer screen displaying a cyanotype by Anna Atkins in a web browser. I can think of few artists whose manner of documenting "nature" encompasses a broader definition of the term, or, for that matter, a more varied means of capturing images.

Lempert came to his approach from the world of science. During the 1980s and early '90s, after studying biology at university, he worked on research projects in Europe, Africa, and on the North Sea, tracking bird populations and authoring academic studies of dragonfly species. At the same time, he collaborated with artists in Hamburg on experimental films and began using a 35mm camera for creative purposes. This hybrid background influences many of Lempert's artistic decisions today. It also makes him a unique figure among contemporary artists: he is as familiar with the ideas of Carl Linnaeus and Charles Darwin as he is with the work of Karl Blossfeldt, Albert Renger-Patzsch, and Bernd and Hilla Becher.

In recent years, Lempert has lavished attention on plant life, often presenting multiple unframed prints in vitrines without descriptive text to encourage associative connections between the densely packed images. (An important aspect of Lempert's artistic practice is the recombining of old and new photographs for each presentation of his work. Like a scientist, he continually seeks new patterns emerging from available information.) In *Untitled (Transmission)* (2014), for example, the shadows of one plant, registered on the fretwork of veins running through a larger leaf, rhyme with the shaft of light in *Untitled (Camera Lucida)* (2018) that, sneaking between branches, falls on an interior door.

Other juxtapositions create looser, even impishly suggestive thoughts. In Lempert's photograph of a morning glory in full bloom, *Untitled (Morning Glory)* (2018), the center of the flower seems to emanate a radiant light. When this picture is paired with *Anna Atkins* and *Untitled (*Arum italicum*)* (2010), two images that involve computer components, the laptop on which I'm writing these words becomes a flowering plant that awakens with me each day. In turn, the flower's concave luminosity reads, alongside the computer screens, as a portal to another world.

Like all of Lempert's work, these pictures offer a combination of specificity—the revelation of a biological phenomenon, the peculiar beauty of a given moment, a wry comment on an artistic predecessor—and a more abstract, lyrical beauty. Taken together, these depictions of reflectiveness, projection, layering, and shadows also become a succinct and potent meditation on photography itself.

Brian Sholis is an independent editor, writer, and curator in Toronto.

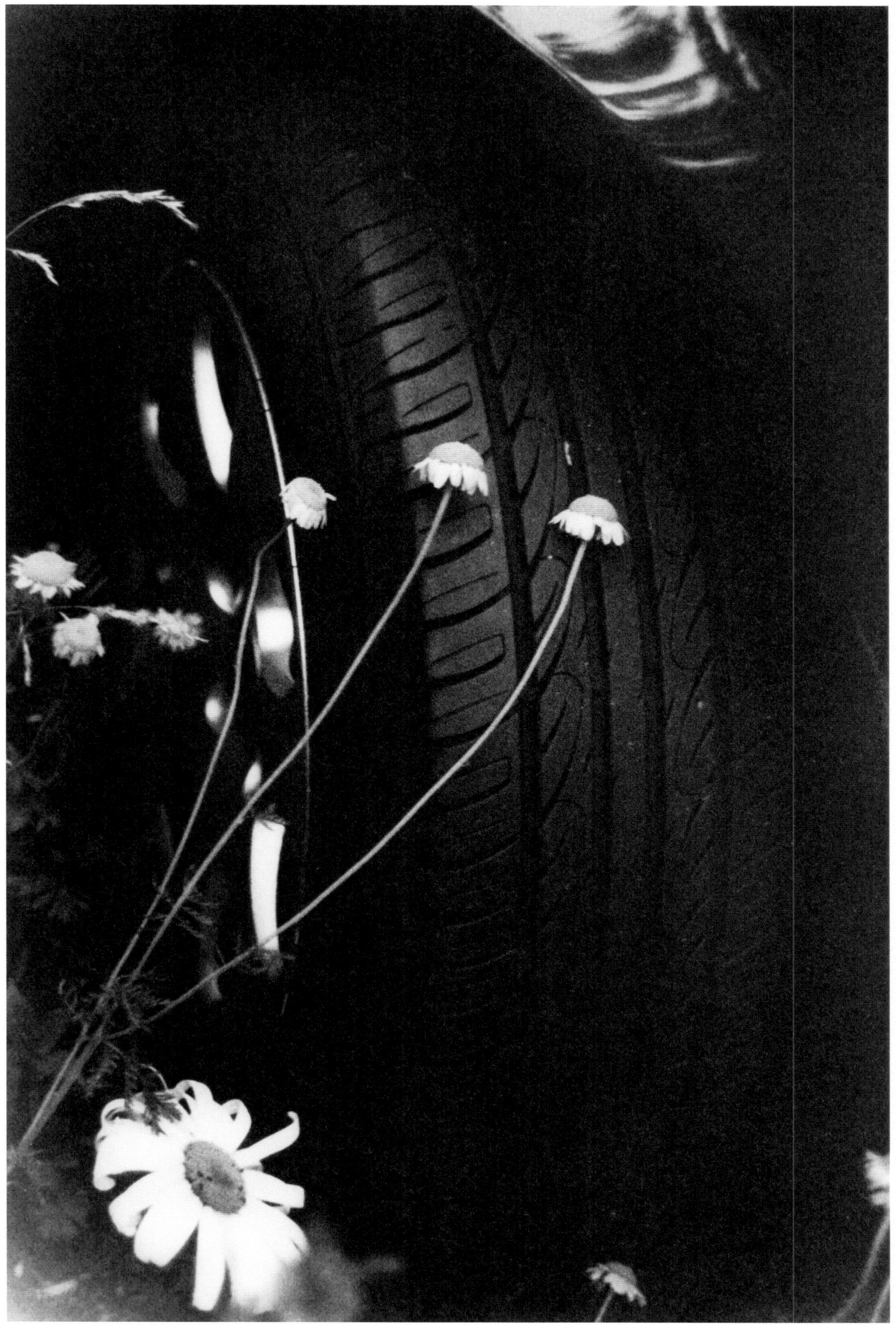

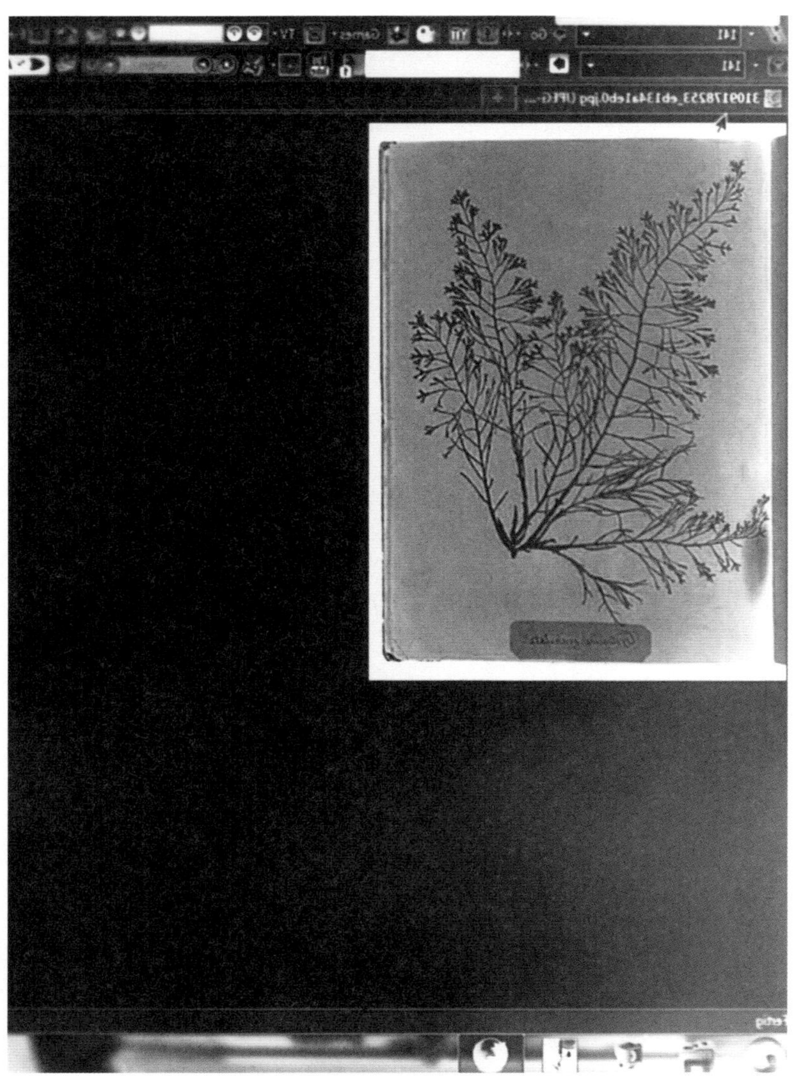

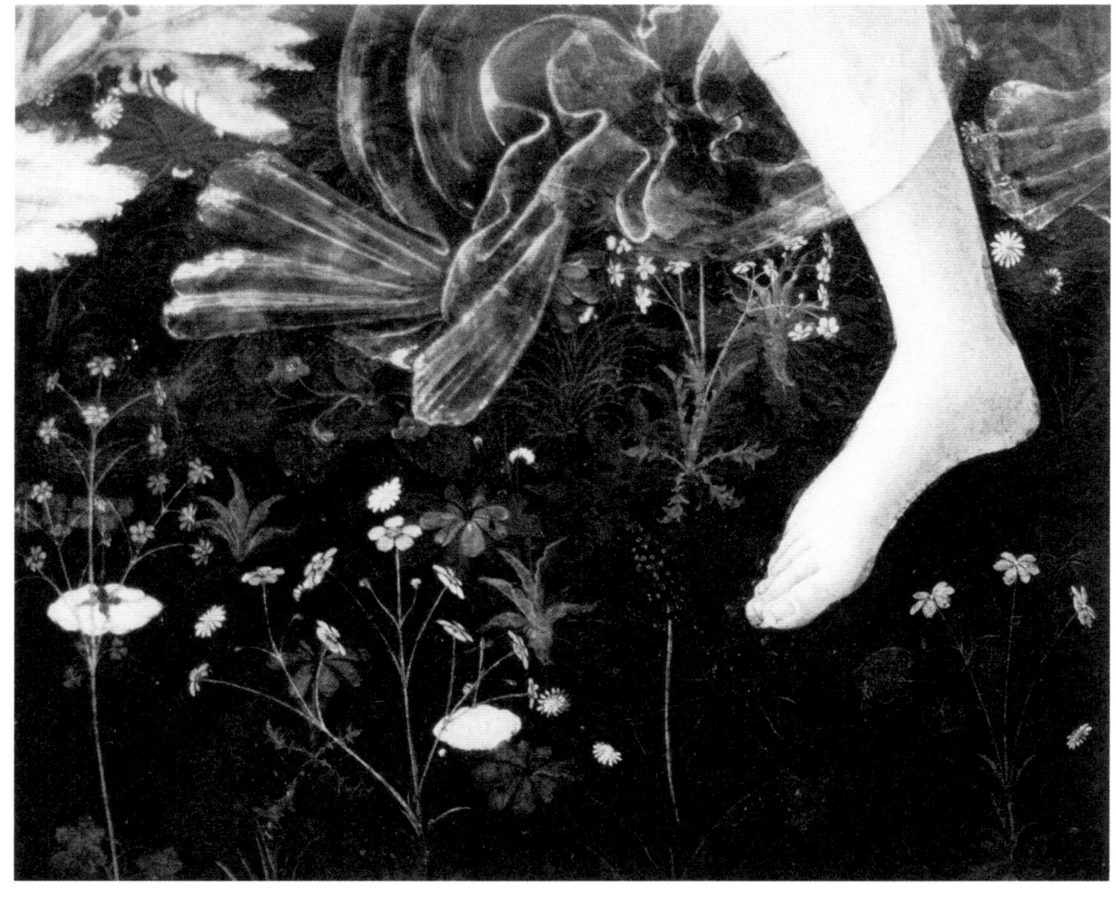

Previous page:
Profile, 2017

This page, top:
Anna Atkins, 2011;
bottom: *Untitled
(Botticelli IV)*, 2018

Opposite:
Untitled (Morning Glory),
2018

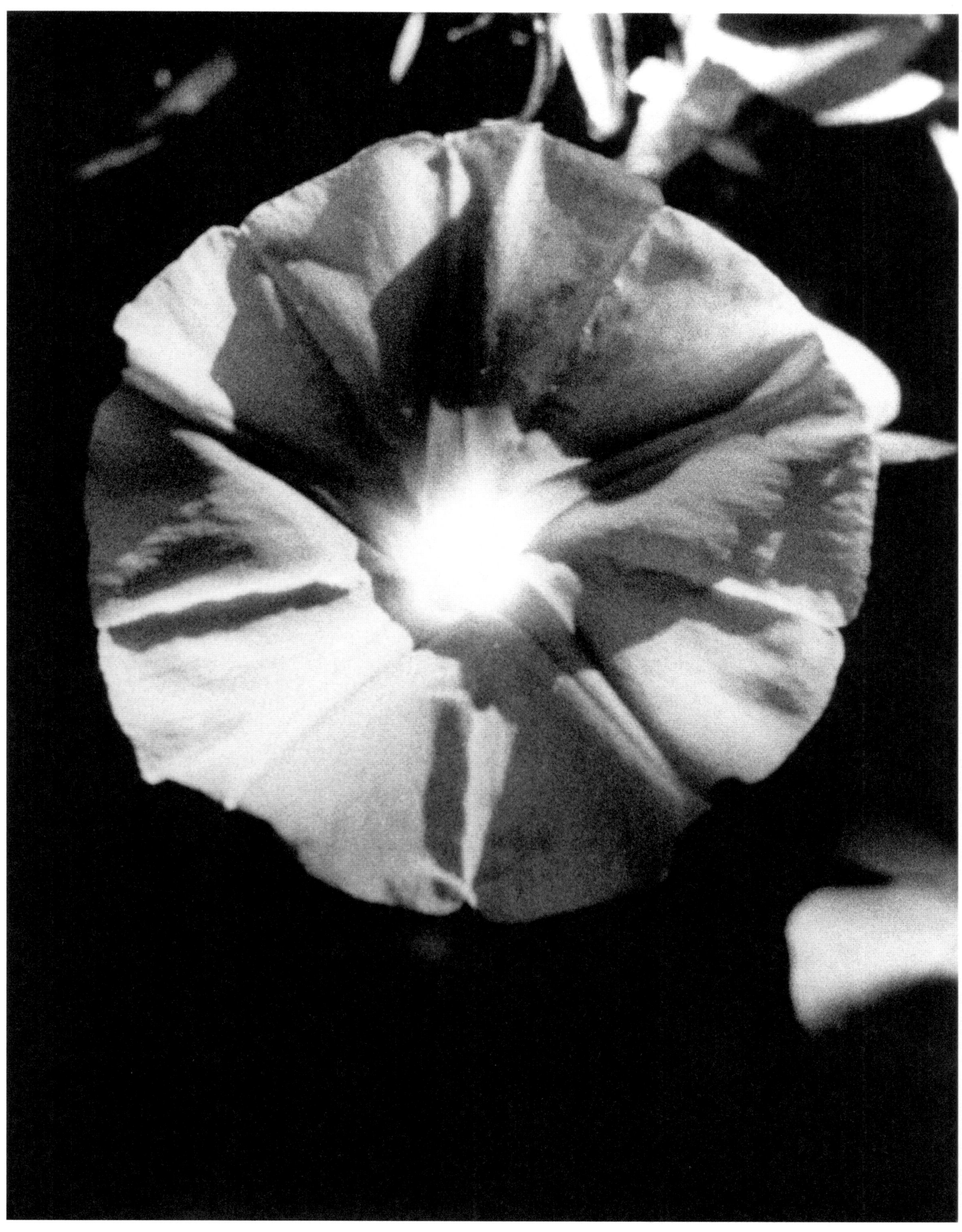

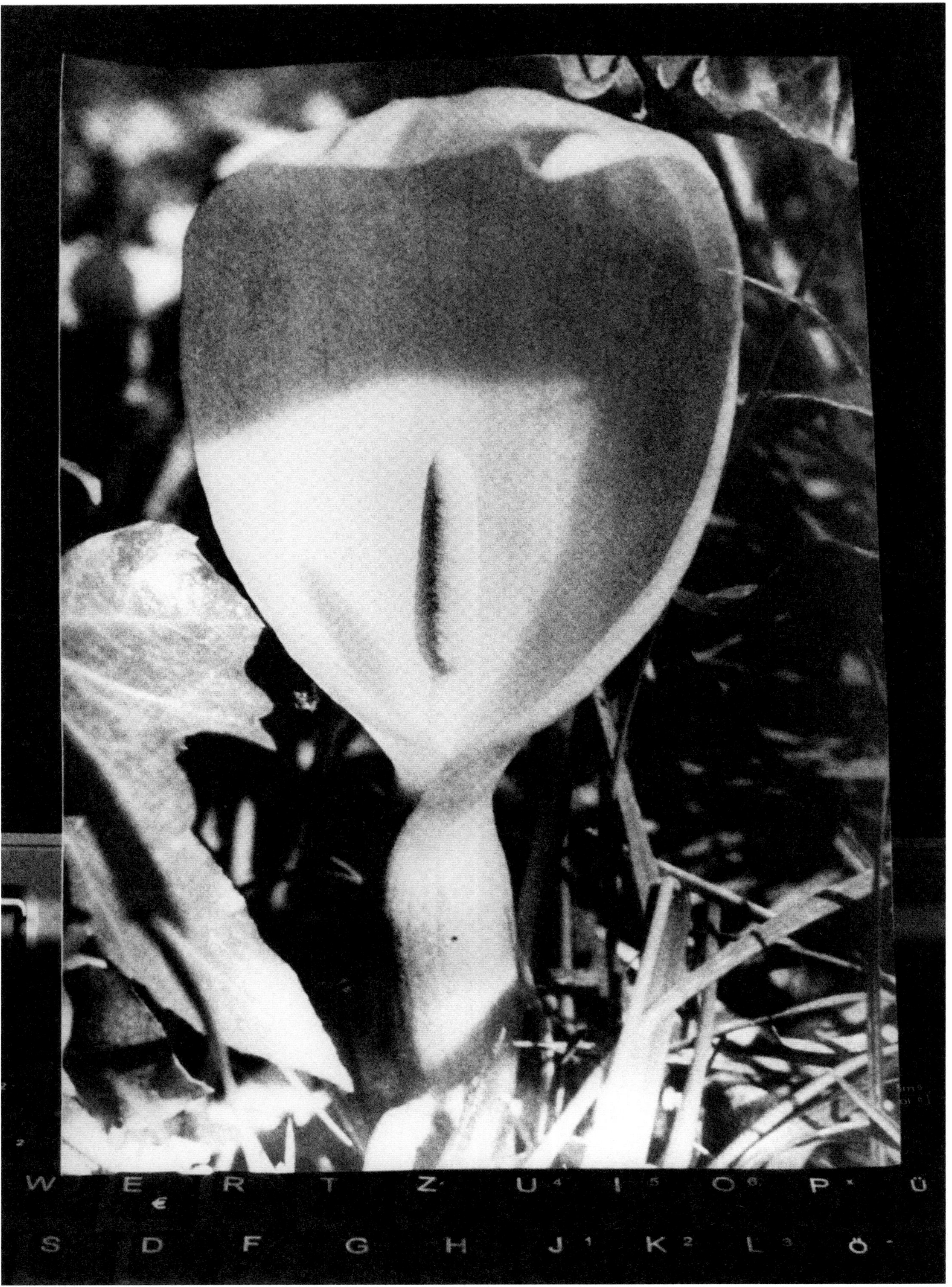

W E R T Z U 4 I 5 O 6 P ' Ü
€
S D F G H J 1 K 2 L 3 Ö -

Opposite:
Untitled (Arum italicum),
2010

This page, top:
Untitled (Camera Lucida),
2018; bottom: *Untitled
(Transmission)*, 2014
All works courtesy the artist

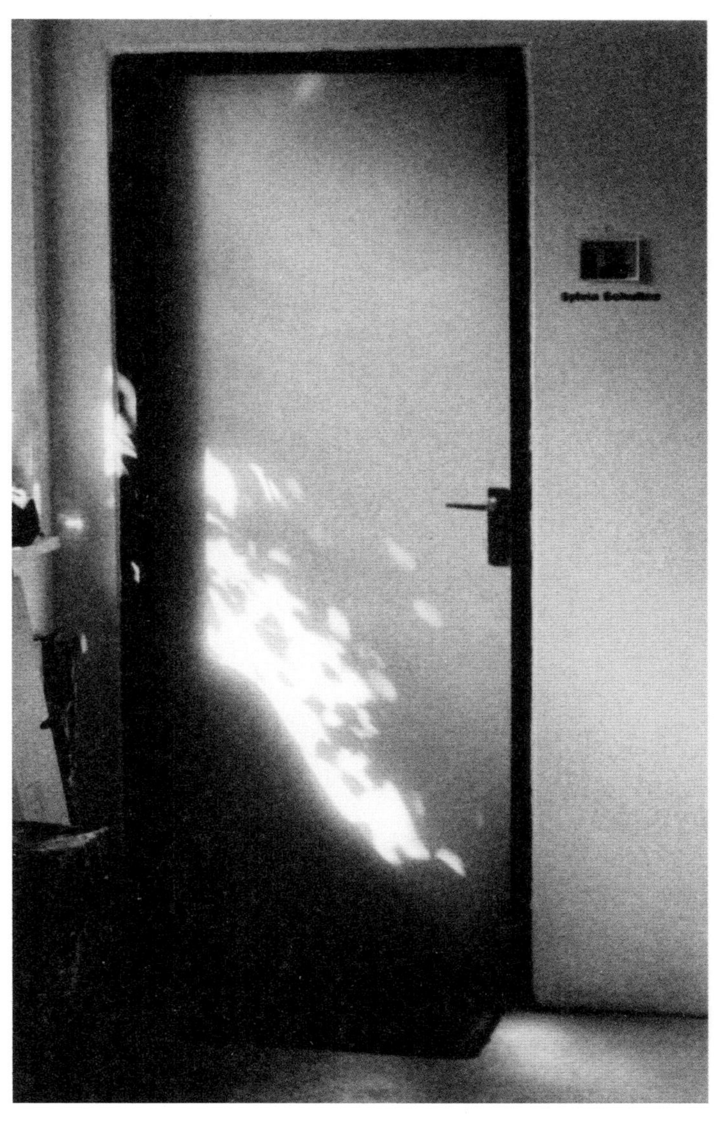

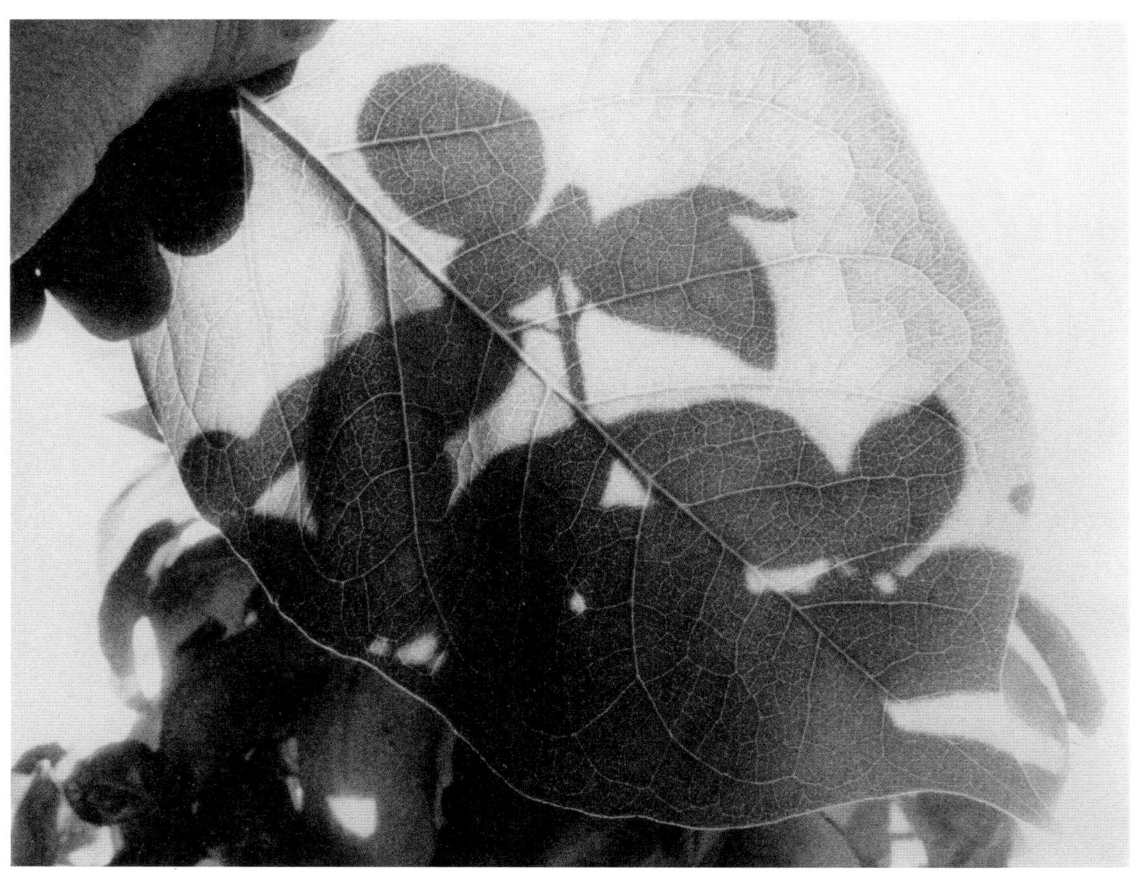

Gideon Mendel

Bronwyn Law-Viljoen

There is a description in J. G. Ballard's novel *The Drowned World* of iguanas perched in the windows of office blocks after a flood inundates London. Bizarre as this image might have seemed in 1962, when the novel was published, such menacing symbols of our disastrous stewardship of the planet are no longer the stuff of postapocalyptic novels, as Gideon Mendel will attest. He has traveled to such scenes in thirteen countries, making photographs for *Drowning World* (2007–ongoing), his project about the global effects of flooding.

Watermarks (2014–18), one of several series in the project, was set in motion when Mendel was in Haiti in 2008. When his two Rolleiflex cameras were accidentally submerged, he continued working, shooting forty rolls of film with cameras that were rusting from the inside out. When Mendel developed the film, he realized that he had, in the strange distortions of the water-damaged negatives, an unexpected materialization of the flood. He had seen such effects before—in the family photographs he often spotted floating by during his forays into flooded neighborhoods.

As a result of this work, Mendel has become an avid watcher of global weather patterns, tracking storms that might produce the conditions in which he prefers to photograph, which have led to profound changes in his work. Because he has no interest in being a storm-chasing equivalent of a war photographer or a producer of aftermath imagery, Mendel has learned to time his trips to flooded zones so that he enters the "drowning world" at its most uncanny: after the immediate catastrophe of the storm and before the receding of the waters. In this eerie calm, when light is reflecting off still water and the sounds of a city have been replaced by bird calls or silence, people feel compelled to return to their submerged homes in order to assess what has been lost, and what might be salvaged.

Mendel sees the flooded home as the logical outcome of a complex chain reaction set in motion by natural events and human shortsightedness: in Kashmir, climate change, environmental mismanagement, and the monsoon; in Brisbane, a rise in ocean temperatures, a La Niña event, and the development of hurricane conditions; in South Carolina, unprecedented rainfall exacerbated by climate change; and in Somerset, damming, deforestation, artificial channeling of rivers, and a massive storm surge.

And if the home is the end point of this fluvial disaster narrative, then photographs from family albums—which Mendel both finds and receives from others—are its most poignant exemplars. Mendel's treatment of these images has evolved over time; he now regards their ordinary materiality as representative of the complex physical and psychological effects of flooding. The photographs become quasi-ethnographic artifacts, their edges curling up and their images obscured by colorful chemical swirls.

Displaying prints from *Watermarks* in the company of other works from *Drowning World*—portraits, films, and photographs of flood marks on buildings—brings loss into sharp focus. They serve as counterpoints to these more consciously composed elements of Mendel's project, their embodiment of private memories mediating the near-biblical symbolism of *Drowning World* and bringing environmental disaster much, much closer to home.

Bronwyn Law-Viljoen is head of the creative writing program at the University of the Witwatersrand, Johannesburg, and editor and cofounder of Fourthwall Books.

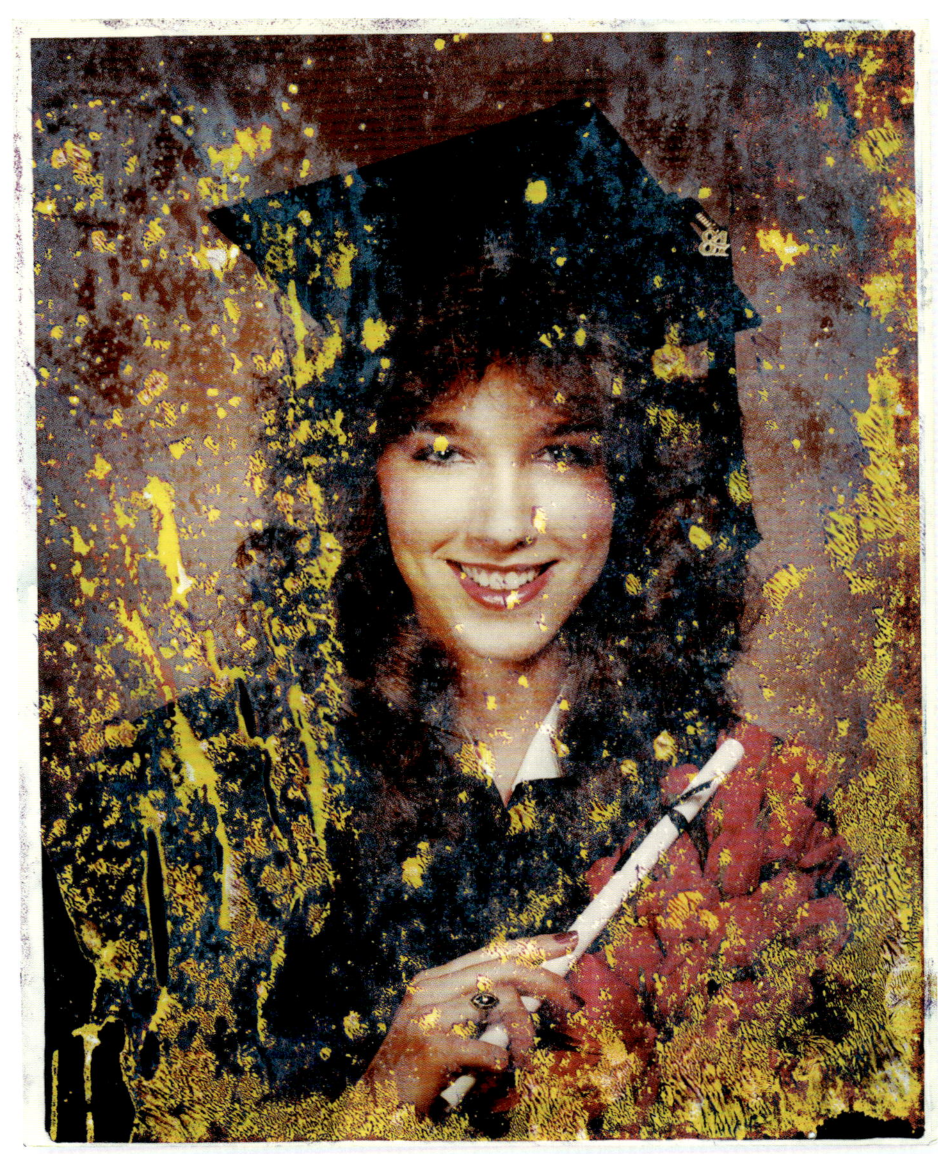

From the home of Belva and Deborah McCormick, Rosewood, Columbia, South Carolina, United States, October 2015

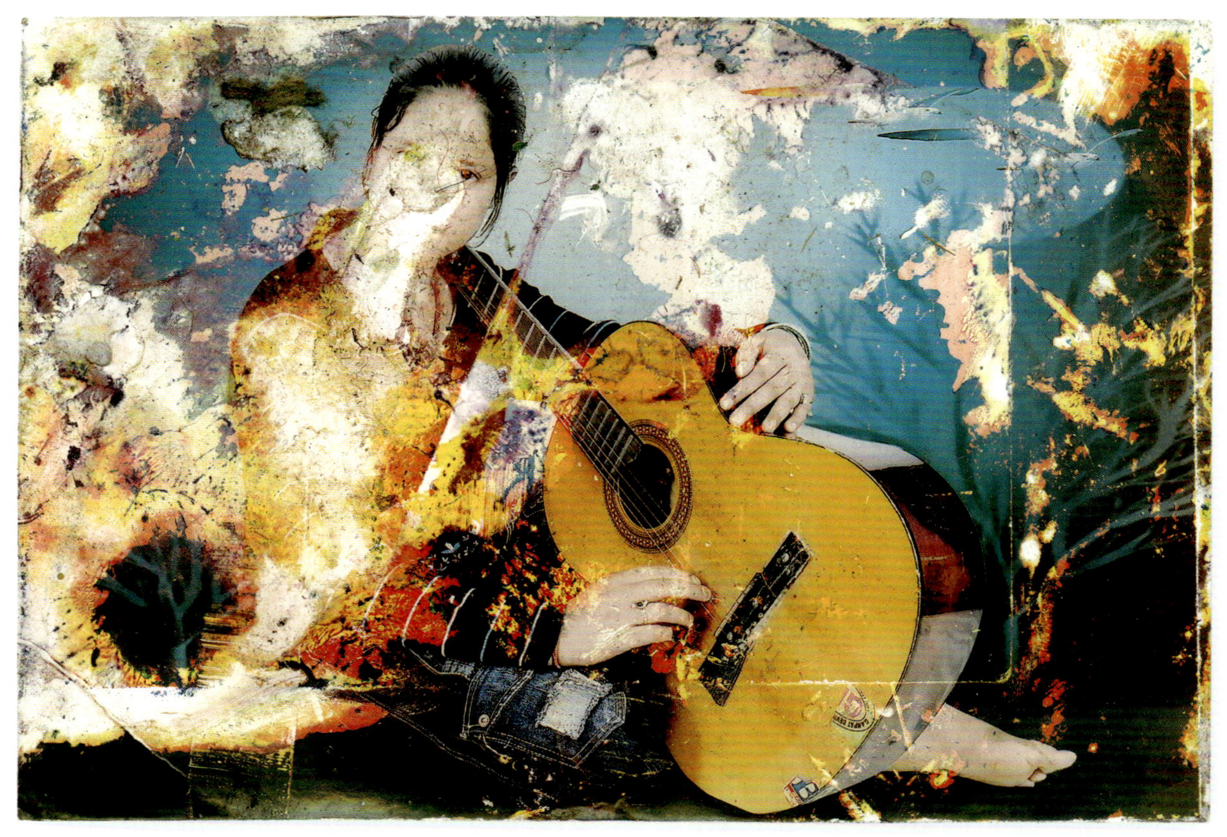

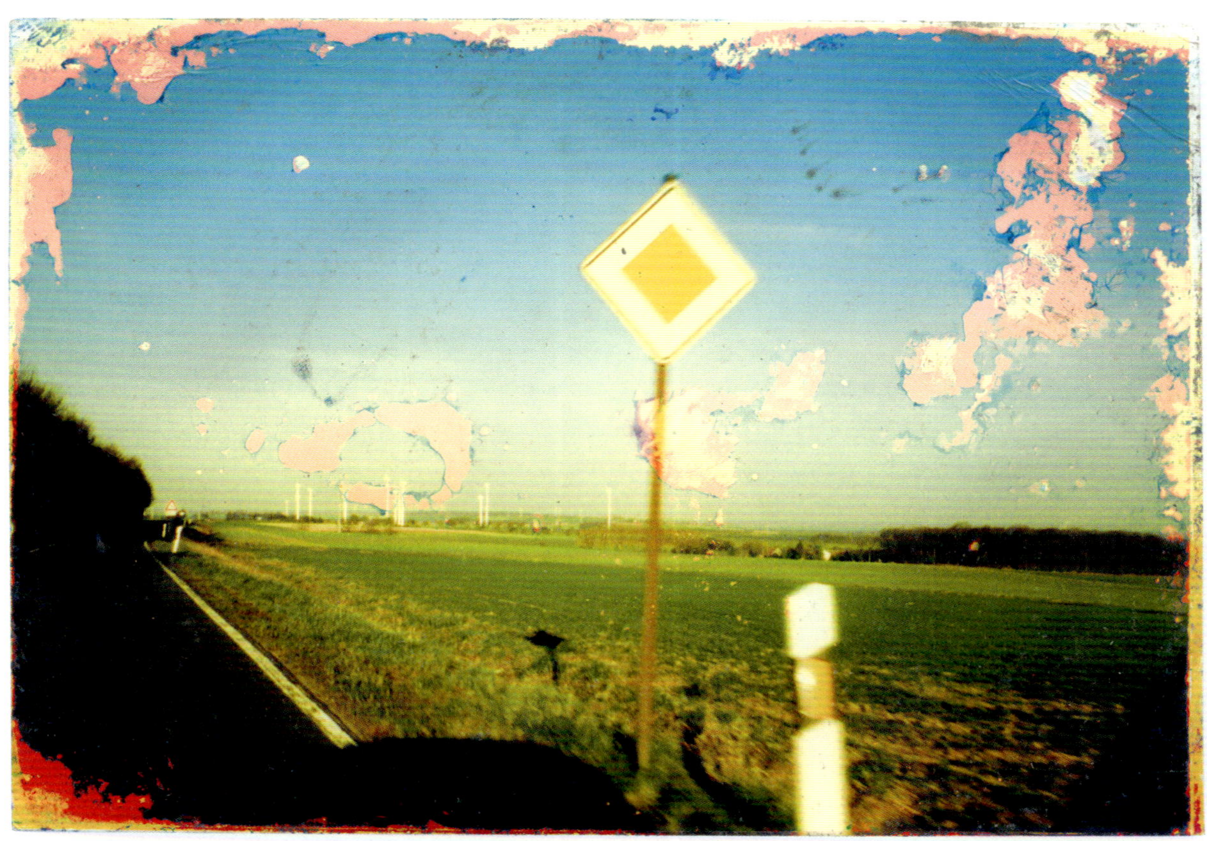

Opposite, top:
From the home of Muskan
and Javed Ahmed,
Mehjoor Nagar, Srinagar,
Kashmir, India, September
2014; bottom: Found on
Goburra Street in Rocklea,
Brisbane, Queensland,
Australia, January 2011

This page:
From the home of Gloria
and Terrence McKeen,
Black Creek, Middleburg,
Florida, United States,
September 2017

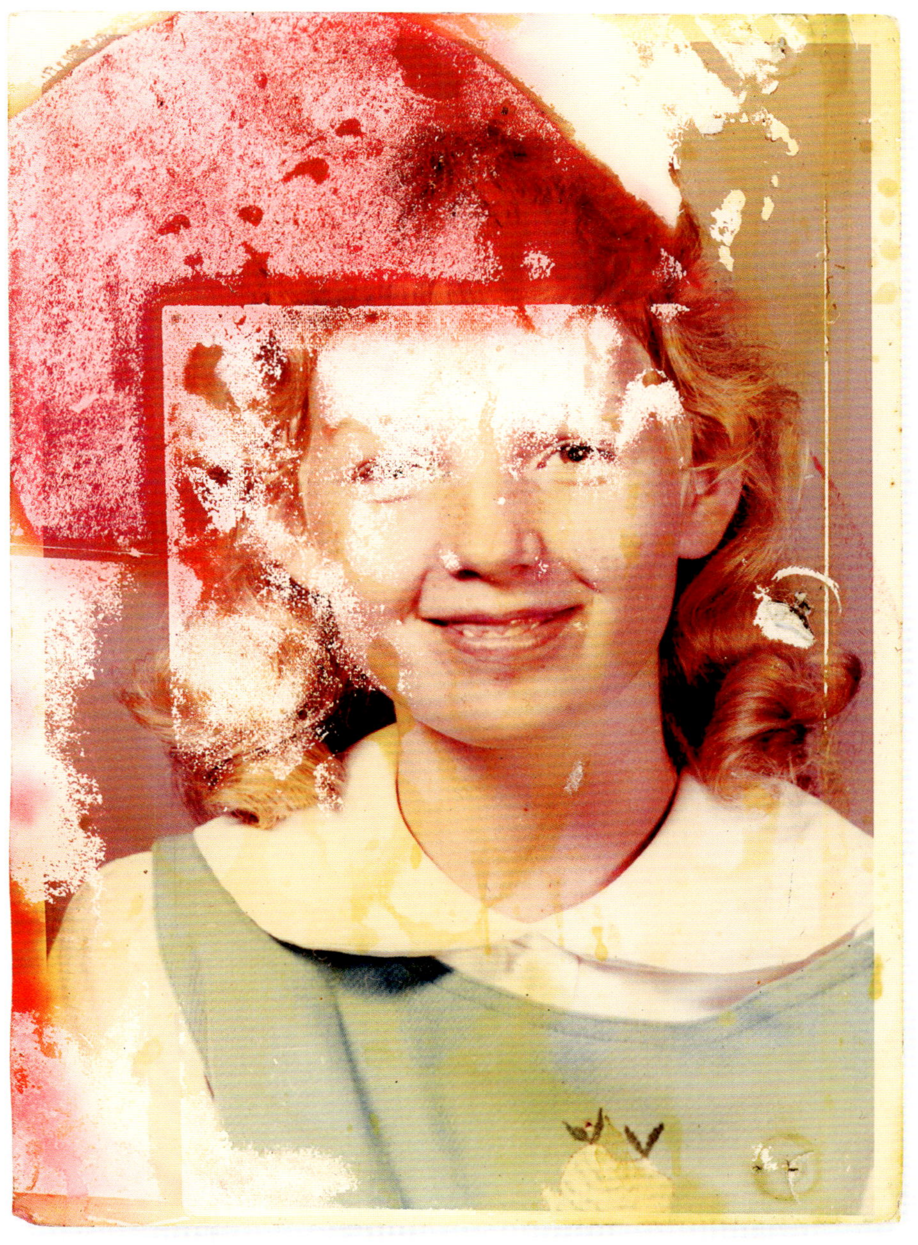

From the home of Muskan
and Javed Ahmed, Mehjoor
Nagar, Srinagar, Kashmir,
India, September 2014

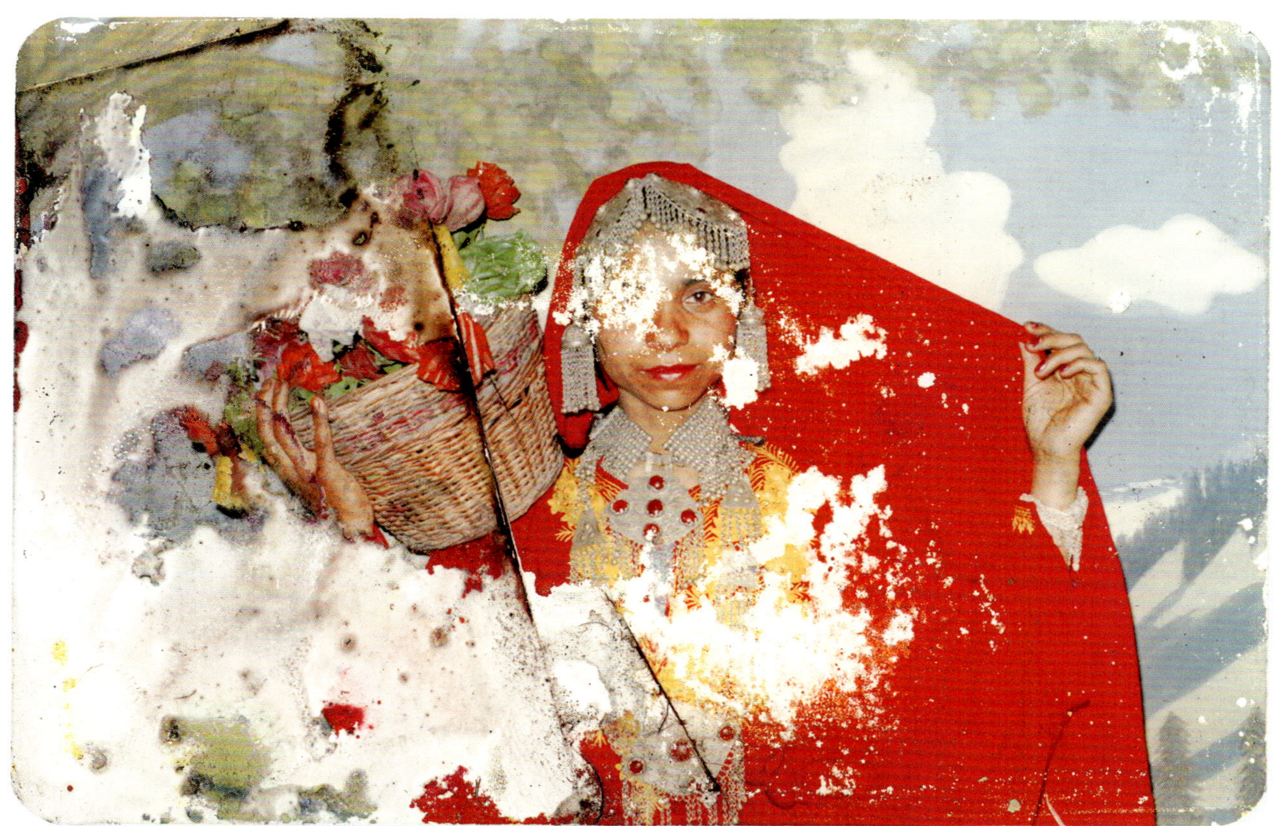

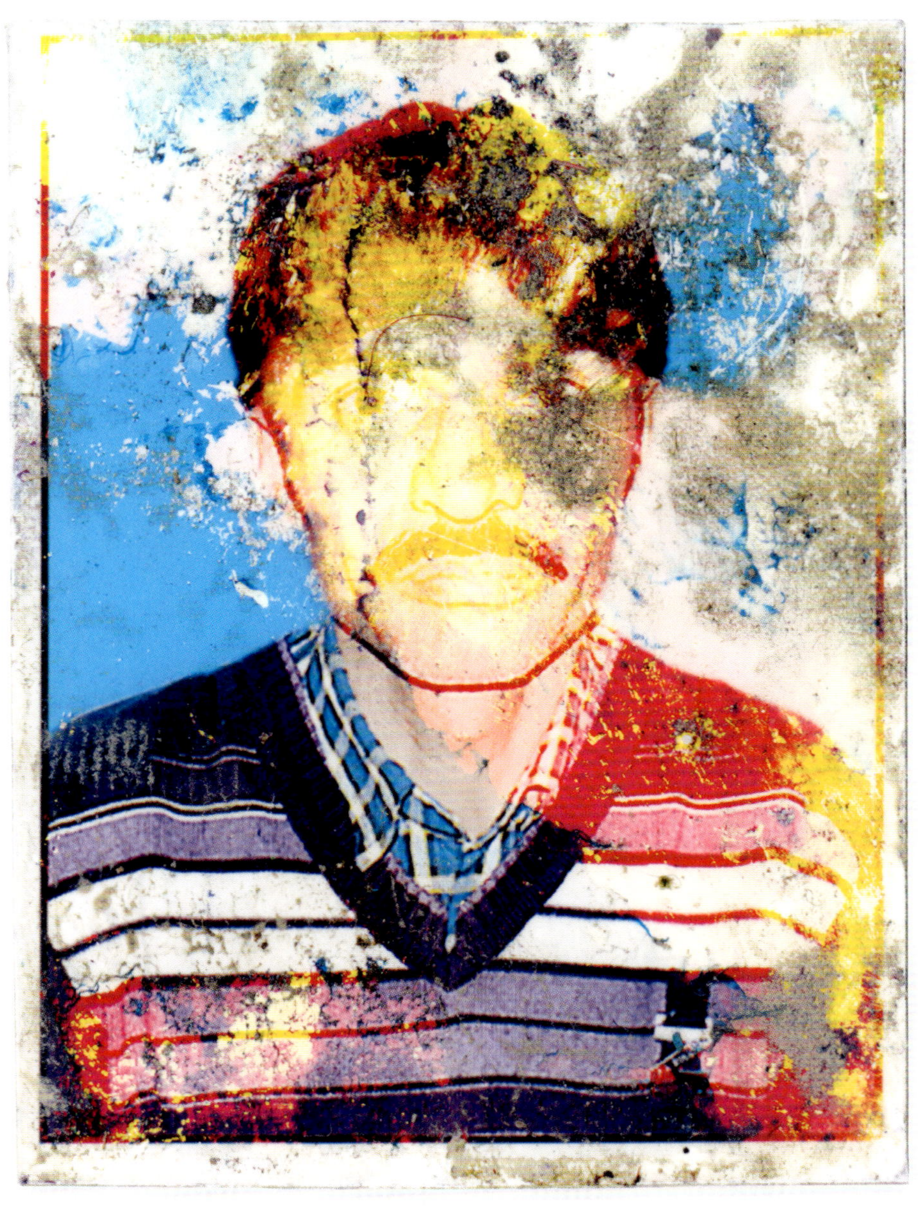

Found floating in floodwater, Jawahar Nagar, Srinagar, Kashmir, India, October 2014

All works from the series *Watermarks* (2014–18), part of the project *Drowning World*
Courtesy the artist and Axis Gallery, New York

Opposite: *Sunday stroll*;
overleaf: *Long stocking,
Split ends*; pages 94–95:
Road trip, Beehive

All photographs from
the series *Plastic Ocean*,
2017–ongoing
Courtesy the artist

From her home in Cape Town, where she's lived for the last six years, Dutch artist Thirza Schaap can walk to the ocean in seven minutes. The waters are wild and cold there, too rough for leisurely swimming. Drawn to them nonetheless, Schaap began walking the beaches with her black-and-white poodle, Iso, and was struck by the perpetual mass of plastic debris washing up on the shore, often caught up in watery strands of seaweed. It always looks, she says warmly, "like there has been a party."

In 2016, Schaap started photographing found plastic sculptures and sharing her pictures on social media. As response to the work grew, she carved out a daily habit, foraging for plastic while out on morning rambles, then going home and making fanciful, pastel-hued arrangements on a table in her garden, and finally photographing her impromptu sculptures. Her lighthearted constructions contain familiar, everyday objects: bottles and lids, balloons, shoes, forks and spoons, toothbrushes, straws, and, of course, the ubiquitous plastic bag. Through Schaap's collaboration with a writer friend, the resulting photographs take on evocative titles: *Sunday stroll*, *Long stocking*, *Beehive*.

Despite their sweet allure, Schaap's images are also deeply troubling. There has, after all, been a global party, and these pictures are glimpses of its ugly aftermath, shards of the unsustainable volume of refuse from our collective voraciousness. As sites of celebration so often appear the morning after, Schaap's compositions are full of spent enjoyment, of things now devoid of use, faded, deflated, or broken. These things have been thrown away, but they persist, unable to decompose, resisting deletion.

Globally, we produce about 340 million tons of plastic each year, and a huge proportion of it ends up in our oceans. There are multiple "great garbage patches" floating languidly around Earth's vast waters, the largest of which, off the coast of Hawaii, holds about 1.8 trillion pieces of plastic and weighs close to 80,000 metric tons. Like a large planetary orb, it continually pulls new objects into its sphere, perpetually accumulating remnants of our modern consumer culture.

Schaap's project *Plastic Ocean* (2017–ongoing) is a whimsical attempt to rescue a few stray fragments from this fate. A habit and a discipline, it has also become a meditative ritual tied to Schaap's deep commitment to living a plastic-free life. Schaap, though, is not keen on pointing fingers or instilling guilt. There's a groundswell now of antiplastic backlash, and Schaap finds inspiration in the community of people working to make things better. As melancholy reminders of the detrimental consequences of our entrenched habits of convenience, her images encourage the possibility, however inconvenient, of changing the way we live for the betterment of the planet and future generations.

Thirza Schaap

Sara Knelman

Sara Knelman is a curator, writer, and the director of Corkin Gallery, Toronto.

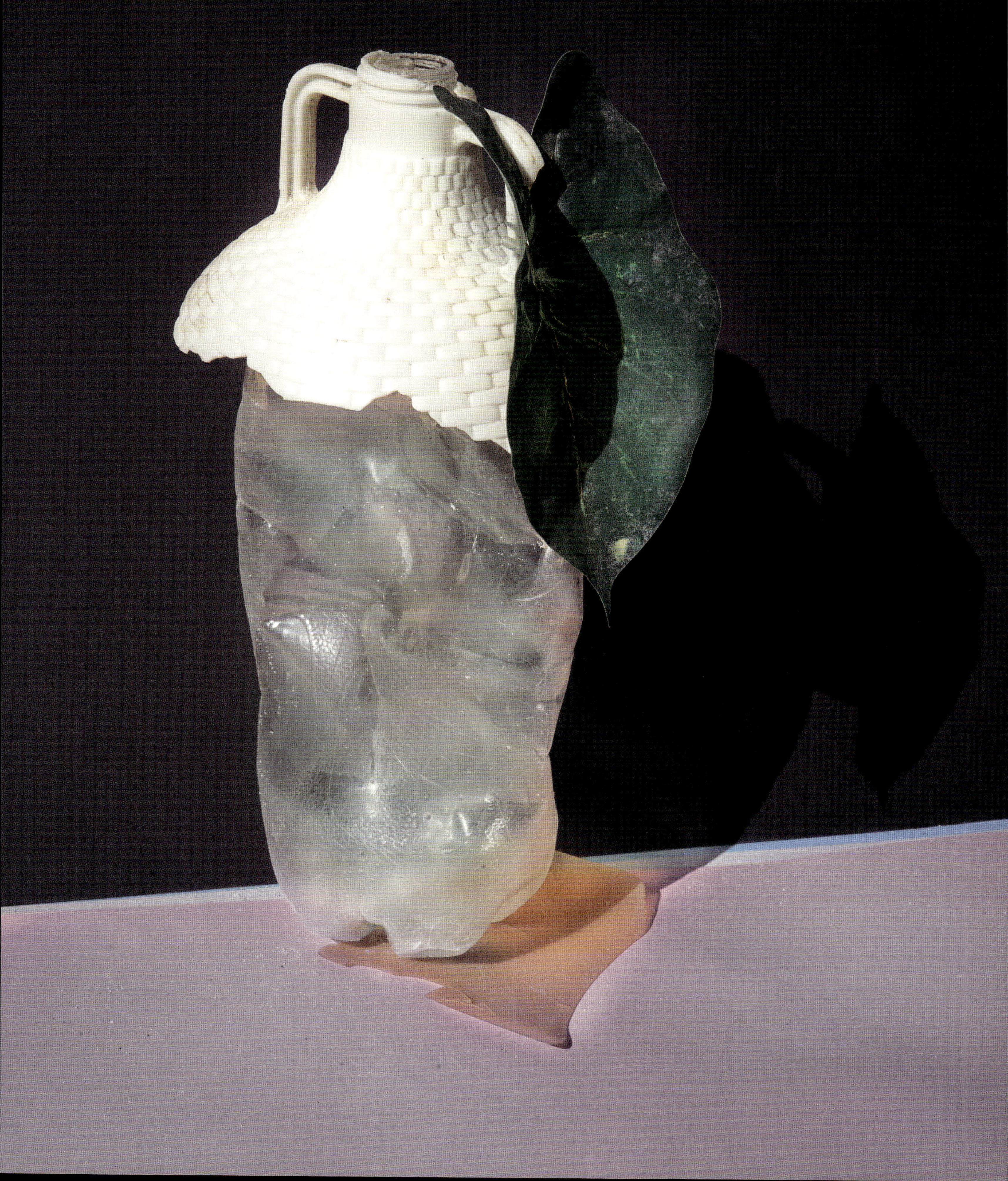

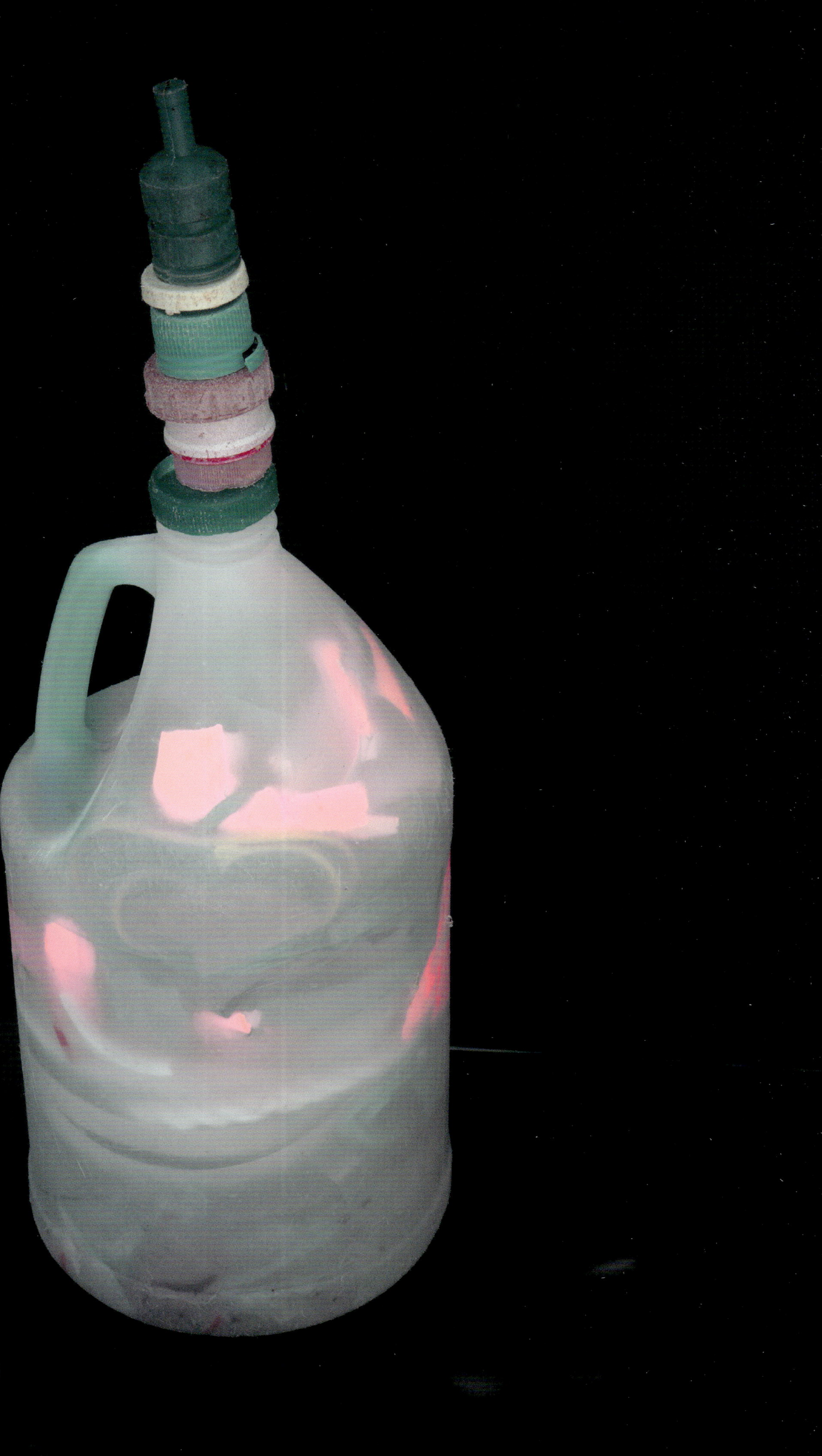

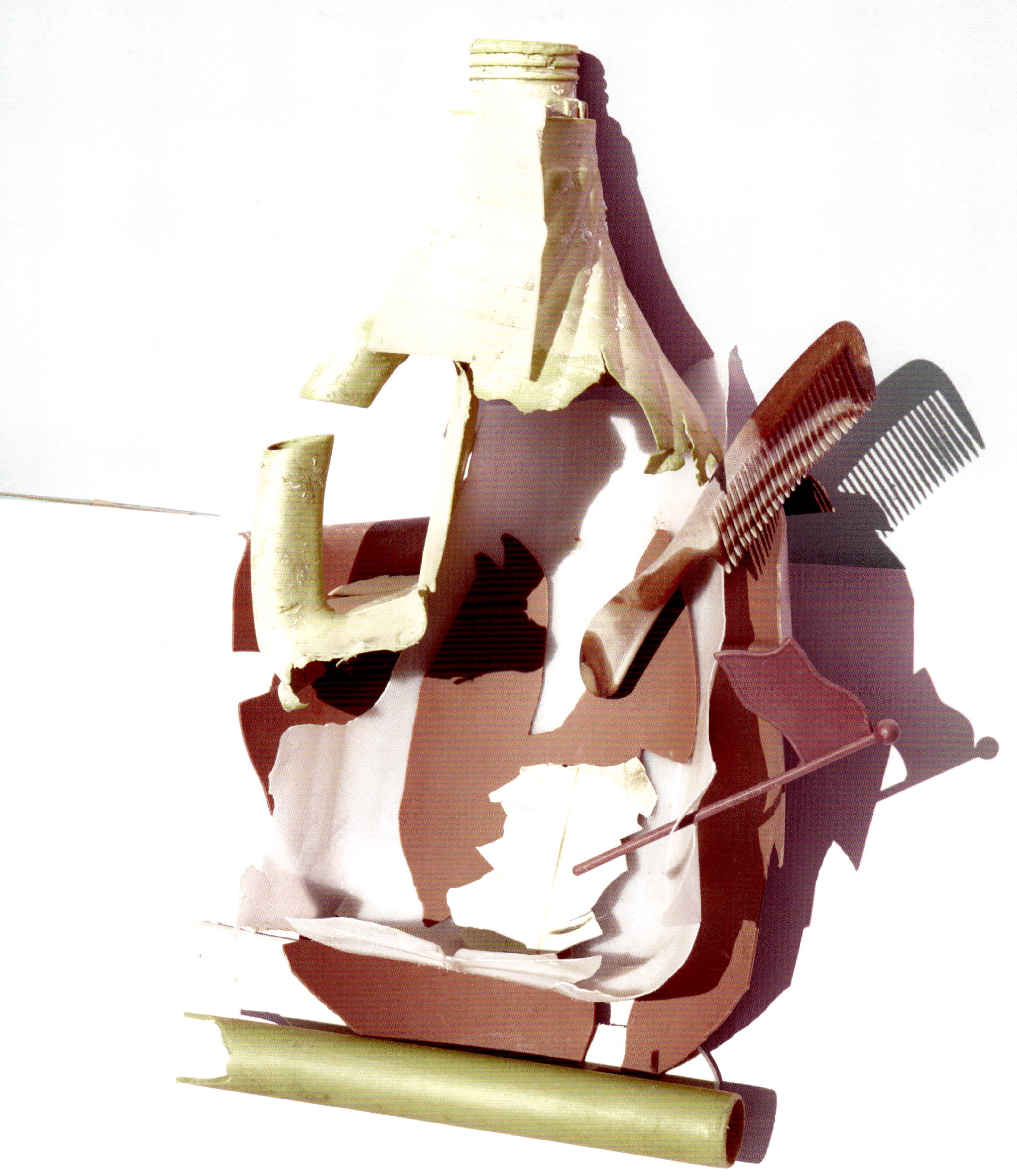

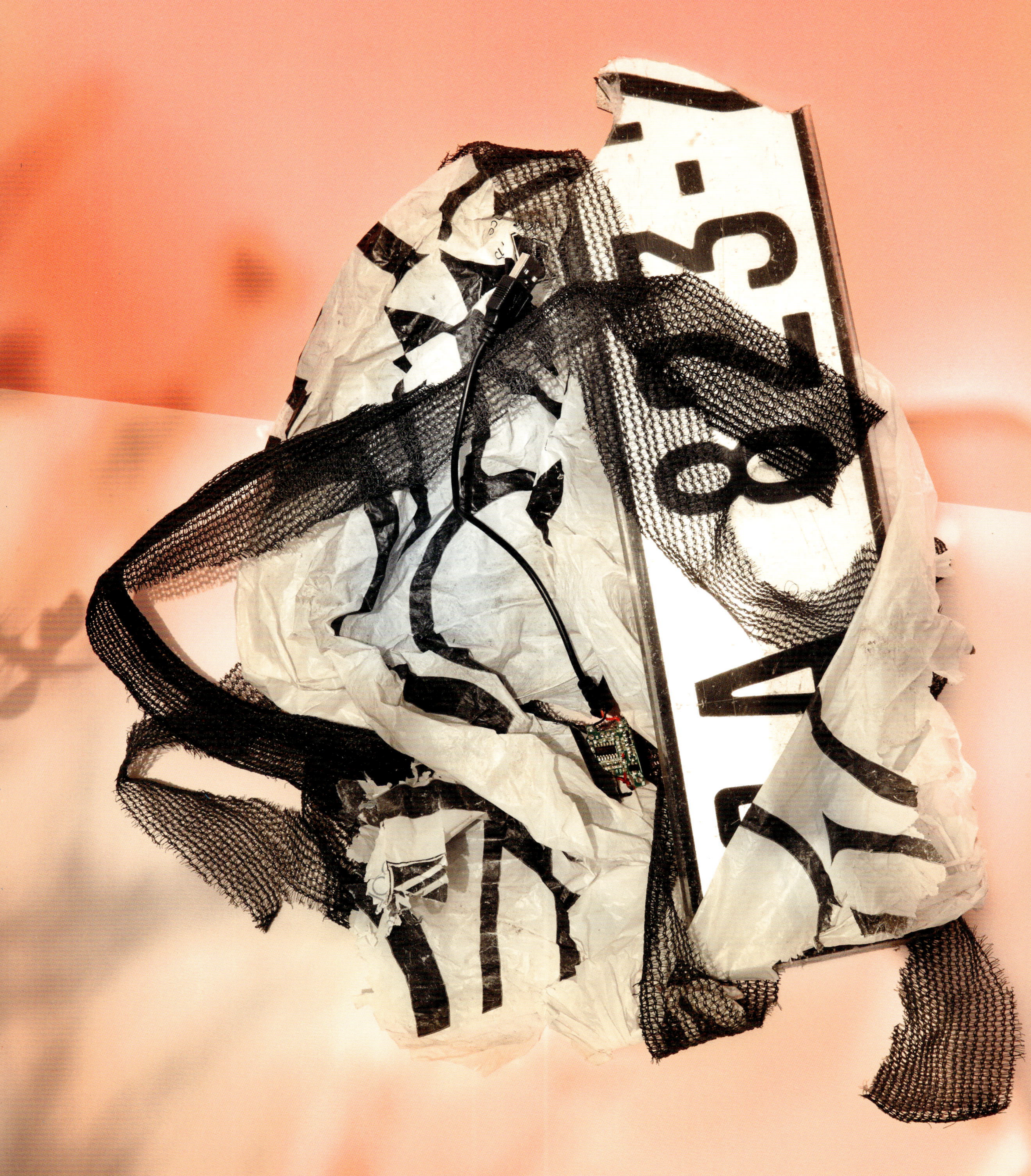

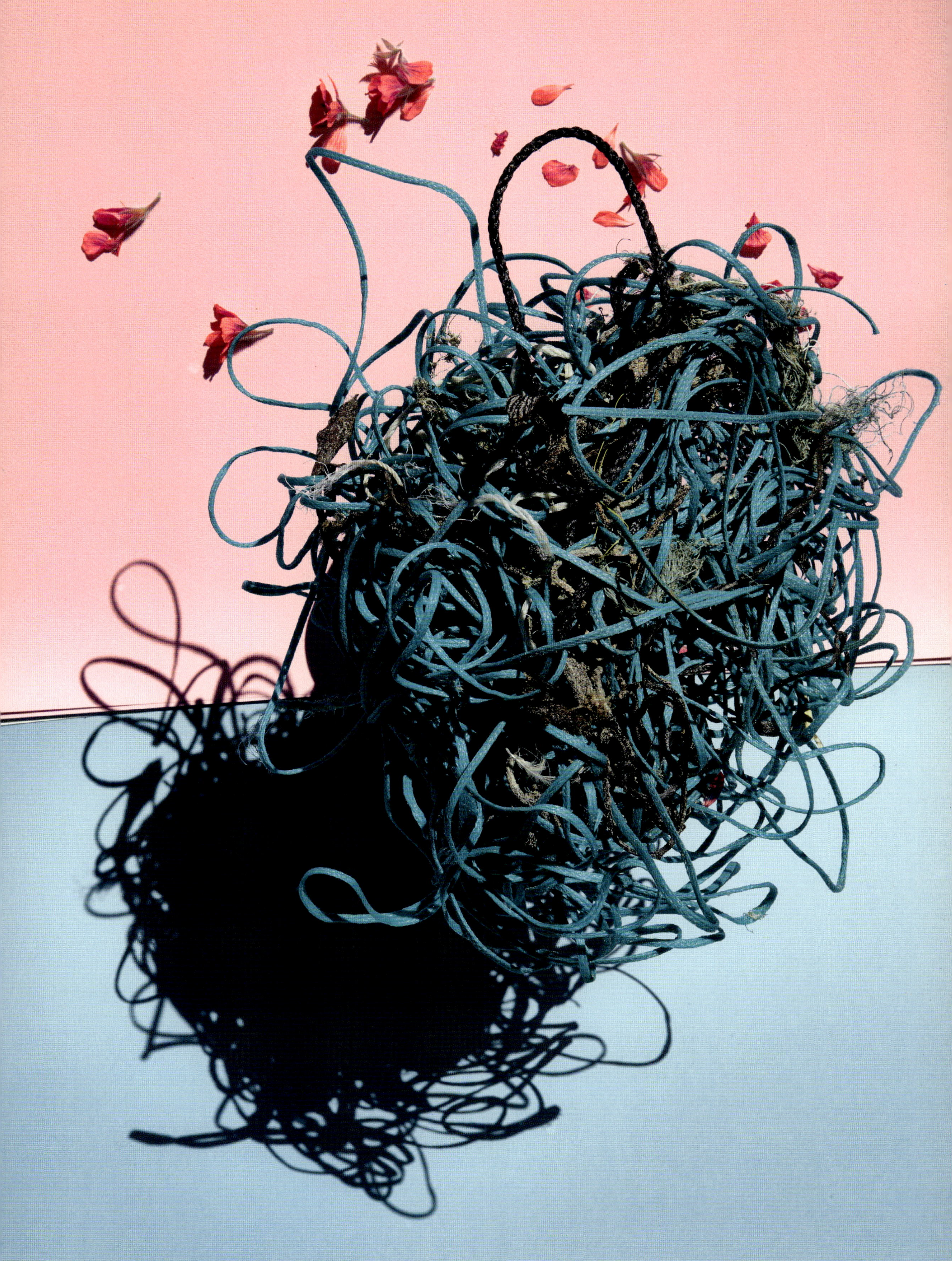

Bruno V. Roels

Brian Dillon

What are billboards if not large-scale collages? Set against the backdrop of a city street, exurban landscape, or empty sky, the billboard opens a mail-slot aperture in reality through which we glimpse other vistas, gleaming bodies, new machines, and gobbets of hectoring or seductive text. As is typical in photomontage, everything is in focus, but the planes of two realms—or more than two, when billboards crowd together—abut each other to flummoxing ends. Seen on the move, at speed, along the edges of a highway, the billboard itself looks mobile—it makes the most mundane journey seem cinematic.

In Bruno V. Roels's series *Fake Billboards* (2018), this sense of the billboard as an icon of modernity (or postmodernity) in motion is still present, up to a point. But there is something more archaic and out of time about these combinations of landscape and blank or obscurely inscribed surfaces. They look like monuments, but to what? Perhaps to an idea of nature: Here is a stand of trees fronted by a "billboard" on which the merest suggestion of a forest has been sketched. Or a distant shore, glimpsed between trees, that turns comically schematic—two lines, one wobbling and one straight—as it crosses the blank rectangular space in the middle of the image. Elsewhere, there are rows of dots or smudges that look like ellipses, as if something has been elided, or is about to transpire, in the surrounding landscape. A billboard with an ellipsis seems to say very simply: *Pay attention*.

Most of all, one notices the palm trees. Roels has explained that they "are not innocent"; they connect biblical stories and imagery to the exotic holiday iconography of the mid-twentieth century, and to the 1980s American television programs and Indiana Jones movies he first saw while growing up in Belgium. They are also a way of introducing another level of repetition into his already reproducible work: in their urban or resort setting, palm trees line up to signal that we are in LA, for example, or the French Riviera, or somewhere that would like to be like those places. These trees are freighted with meaning, but they are also abstract explosions or sculptural forms, and, in some of Roels's photographs, they seem to endlessly ramify, spreading into the slightly bullied corners of the black-and-white image, as if they might eventually darken its entire surface.

In J. G. Ballard's short story "News from the Sun" (1981), as so frequently in his fiction, technology has effected some transformation of human perception—in this case, linked to humanity's forays into outer space. The human eye has come precisely to resemble a camera, recording a series of separate images. The natural and artificial have become ruinously involved and can no longer be separated. "Seen from the speeding car, the few frayed palm trees along the road had multiplied themselves.... The lakes had been the multiplied images of the water in that tepid motel pool, and the blue streams were the engine coolant running from the radiator of his overturned car." With their simultaneous embrace of venerable analog practices, conceptual repetition, and enigmatic montage, Roels's *Fake Billboards* proposes a comparable collision of categories—and an eerie emptiness. In one picture, the palm trees have been interrupted by a "billboard" on which a dozen small circles, irregularly spaced, suggest a diagram of a solar system, and pale infinite space.

Brian Dillon is the U.K. editor of *Cabinet* magazine and the author, most recently, of *Essayism: On Form, Feeling, and Nonfiction* (2018).

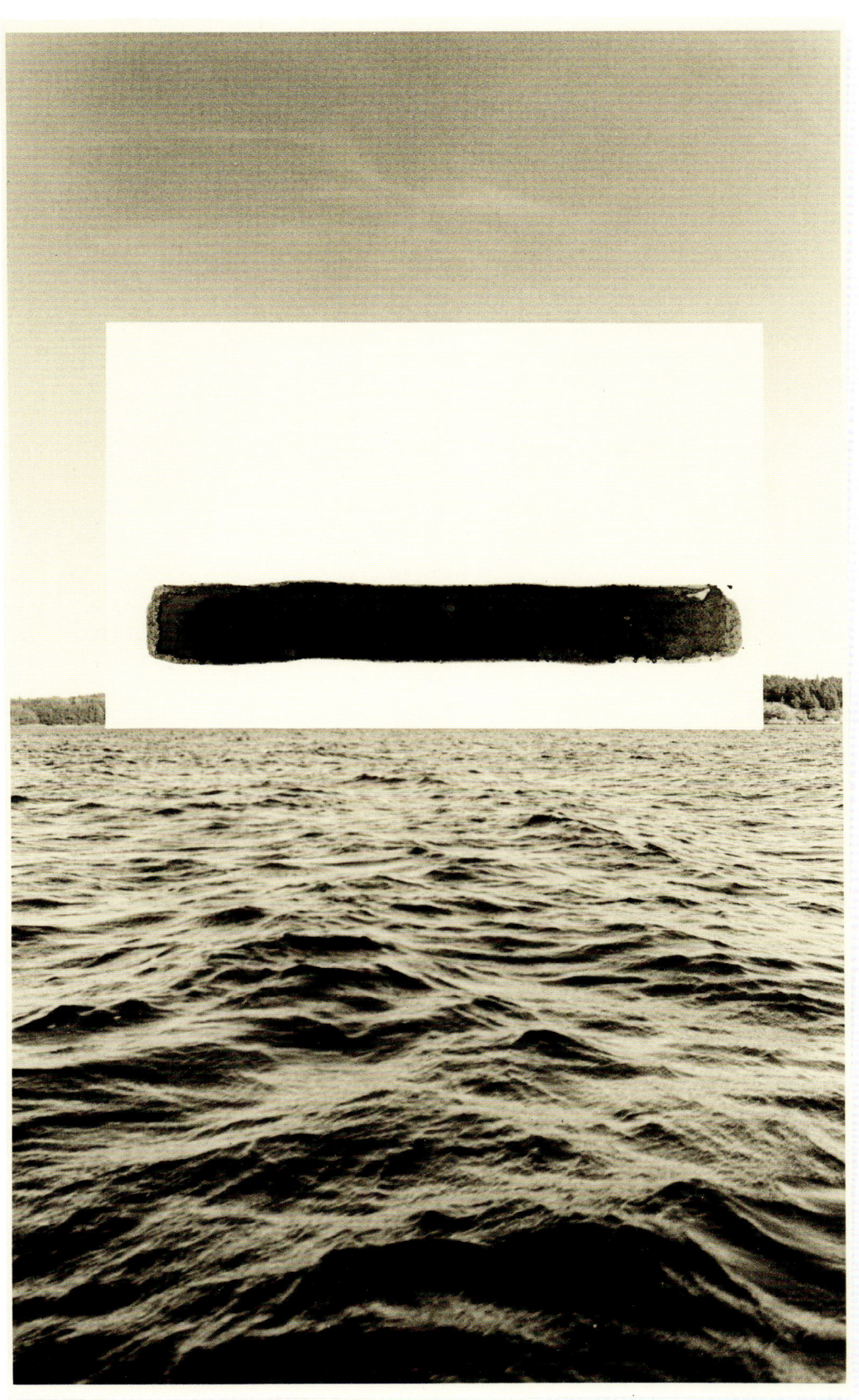

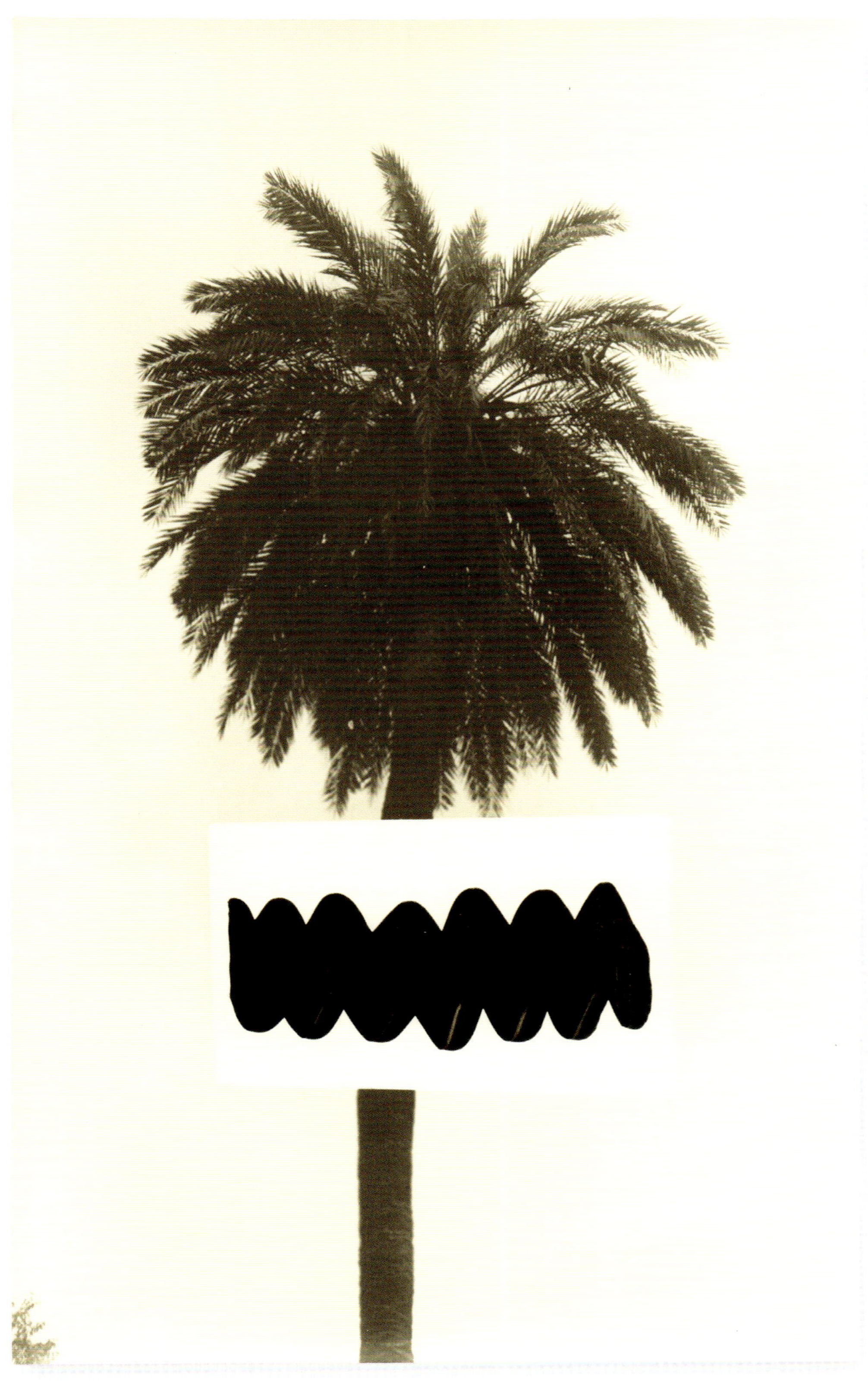

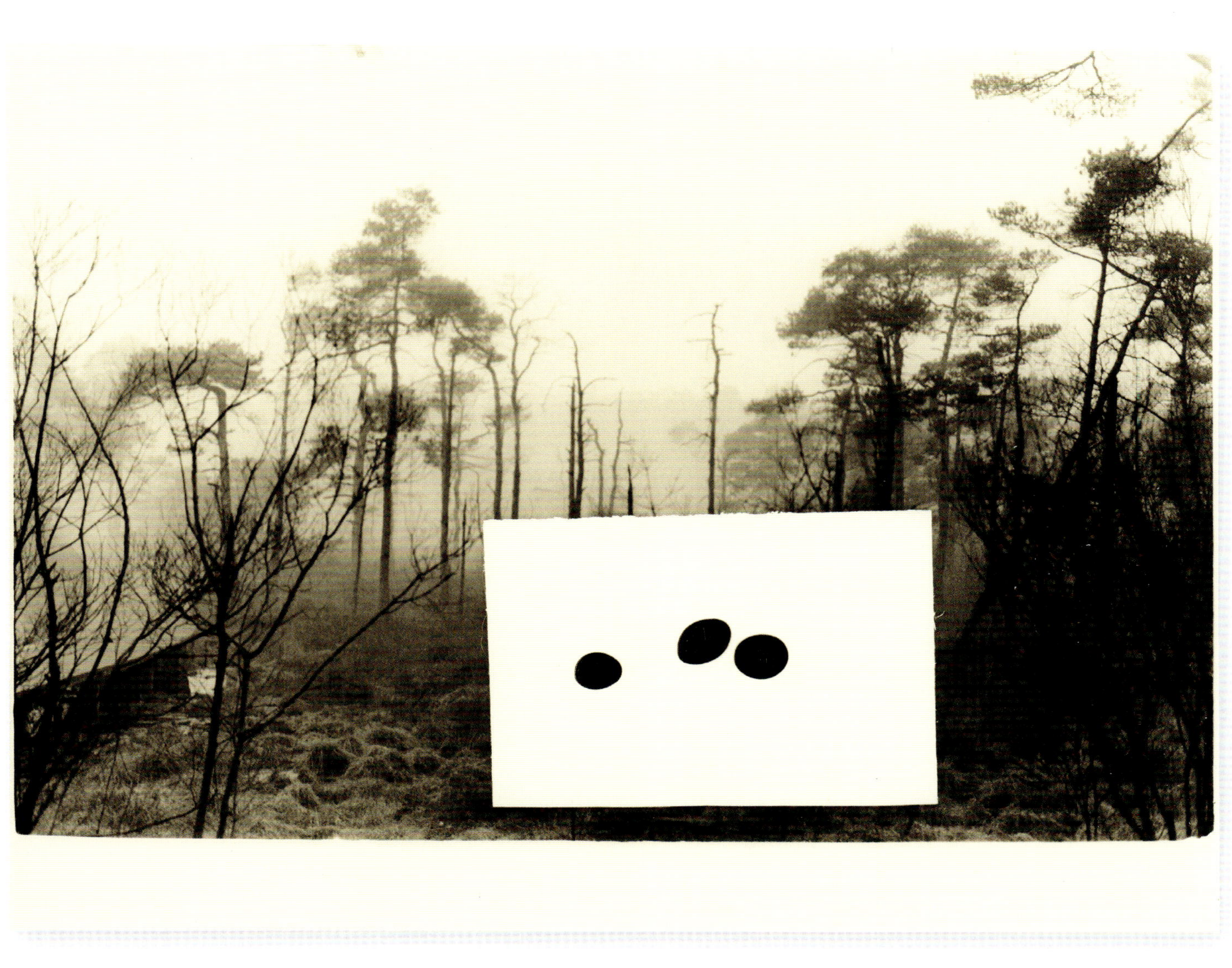

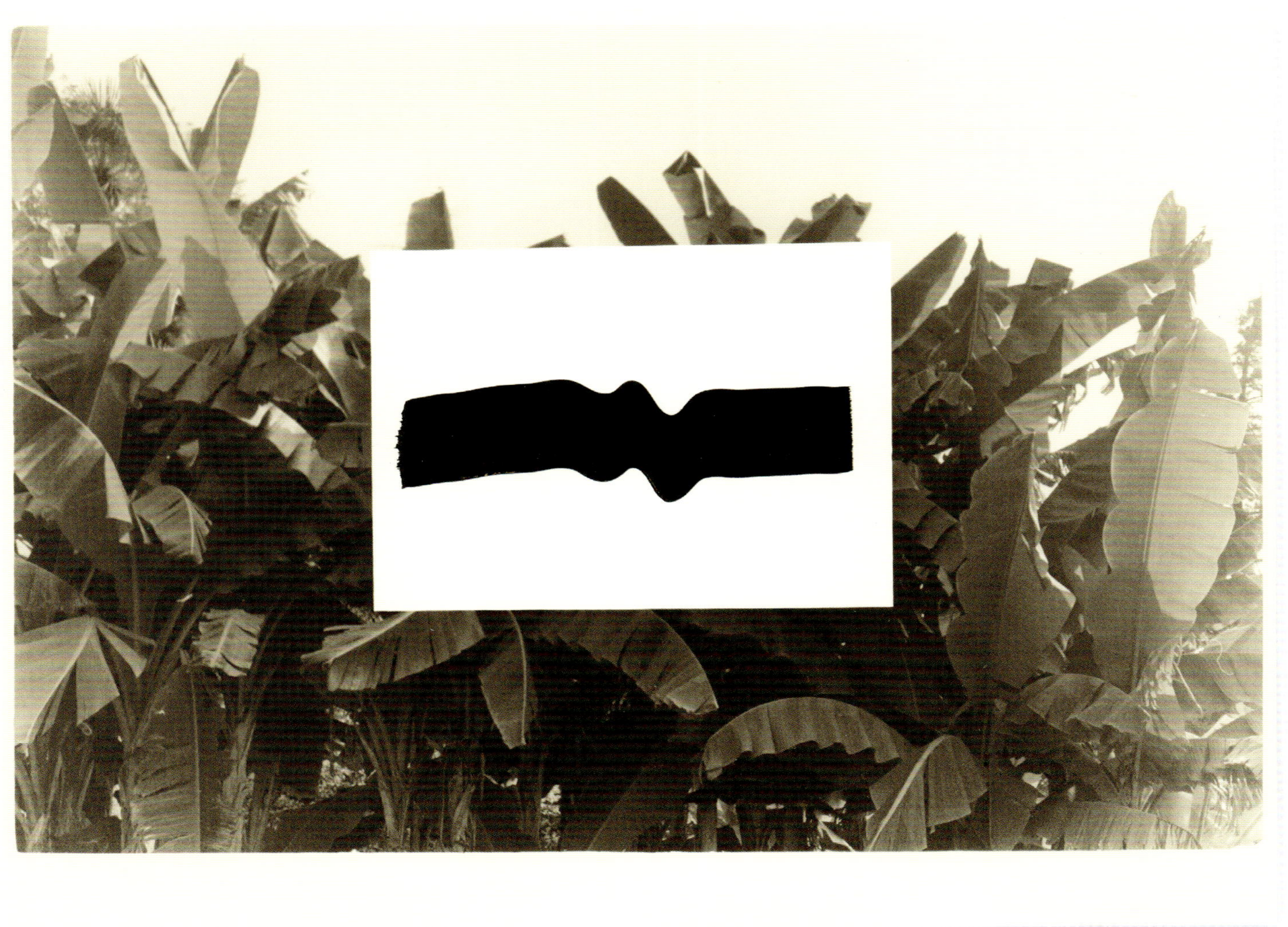

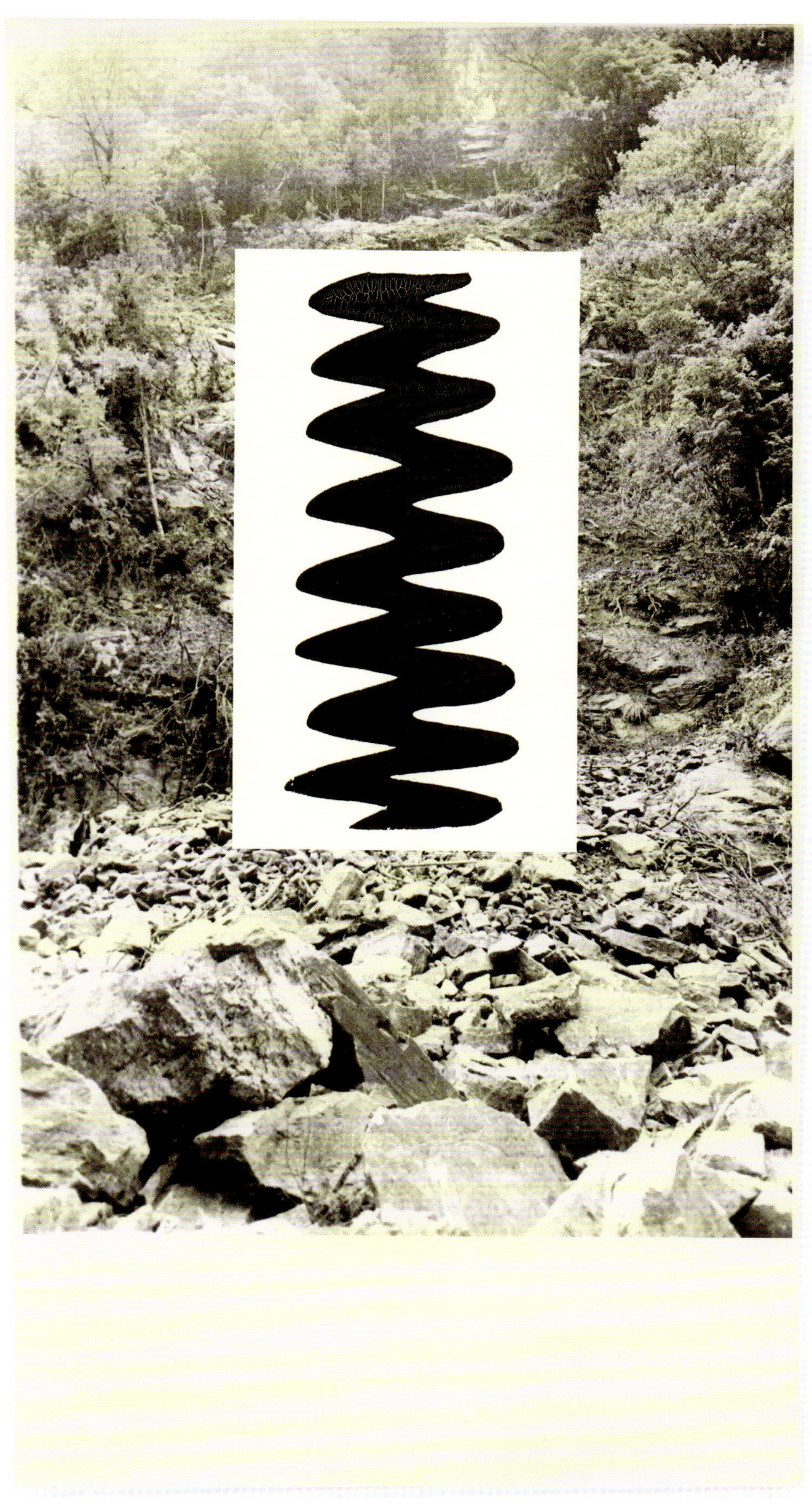

All works from the series
Fake Billboards, **2018**
Courtesy the artist and
Gallery Fifty One, Antwerp

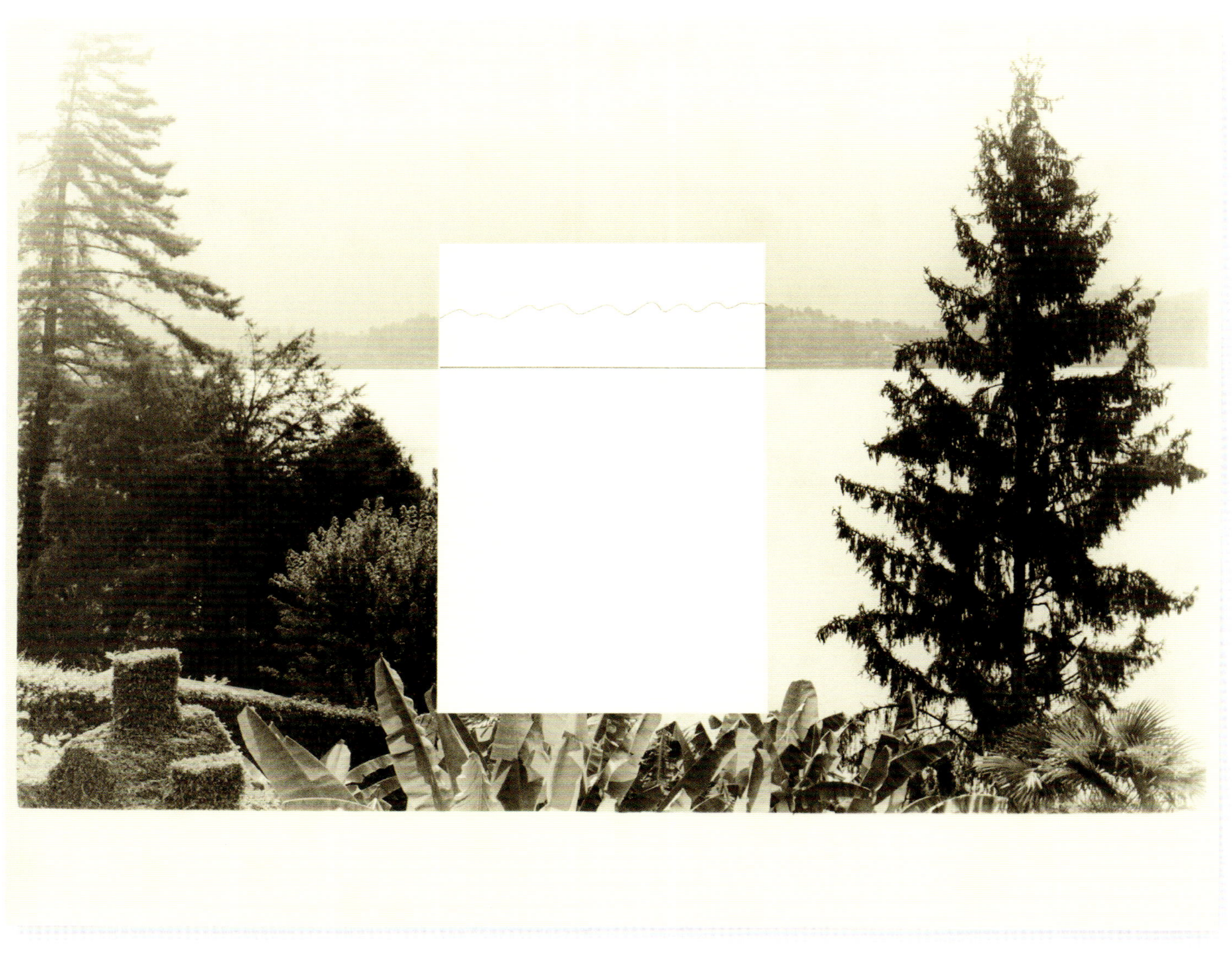

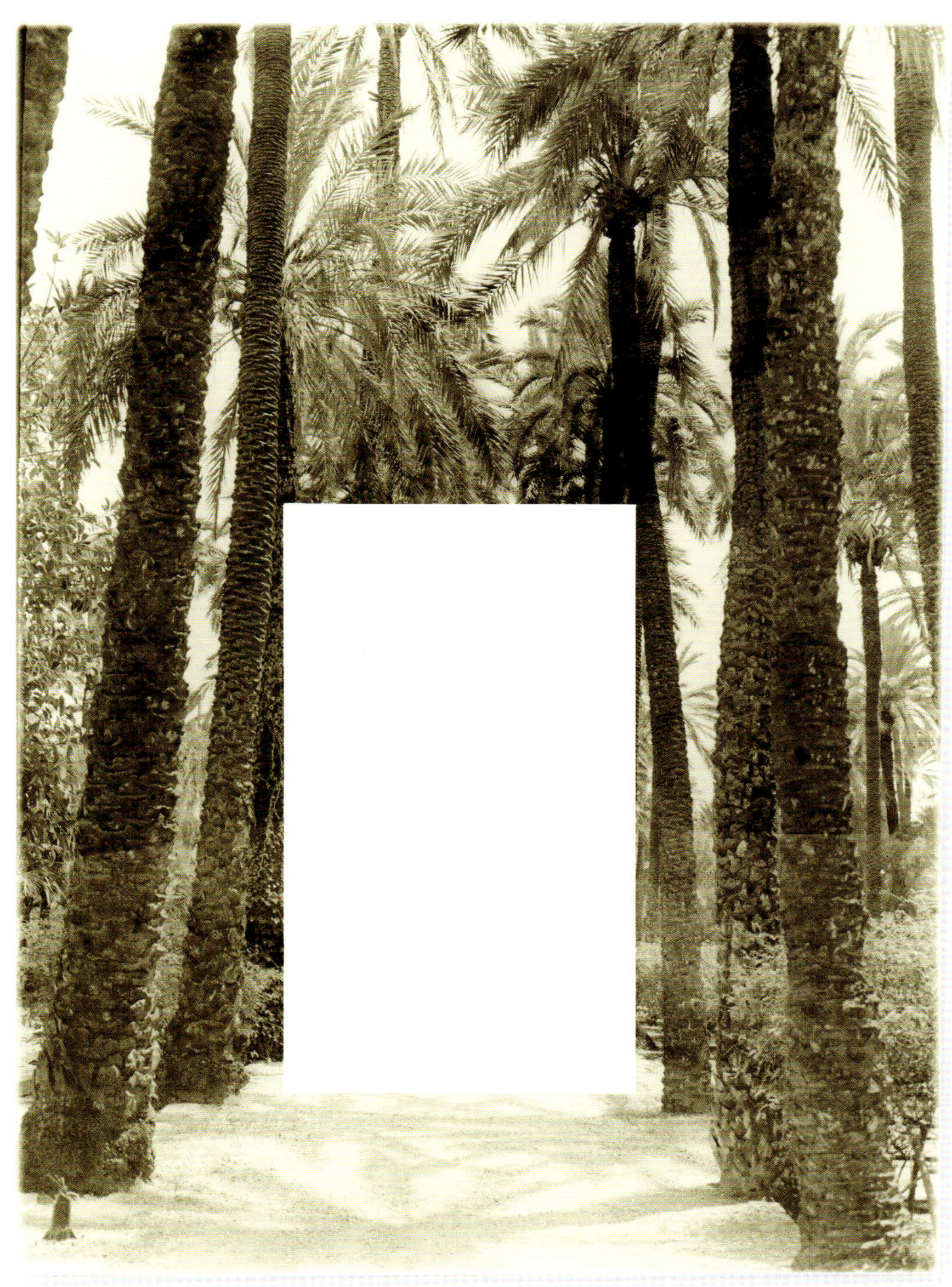

LIQUID

Arguiñe Escandón & Yann Gross

ROADS

Emmanuel Iduma

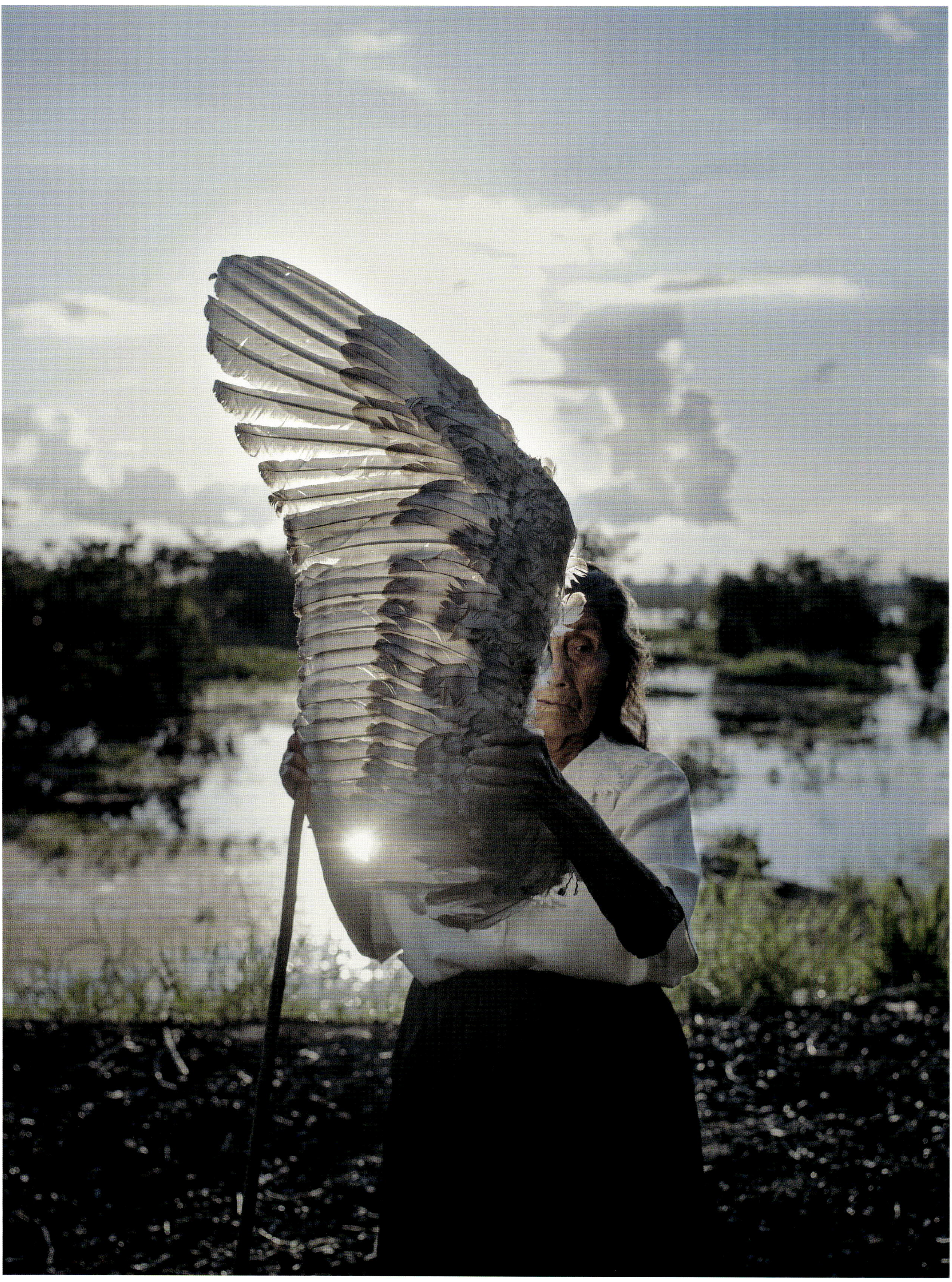

All works from the series
Tamamuri, 2018–ongoing
Courtesy the artists and
Wilde, Geneva

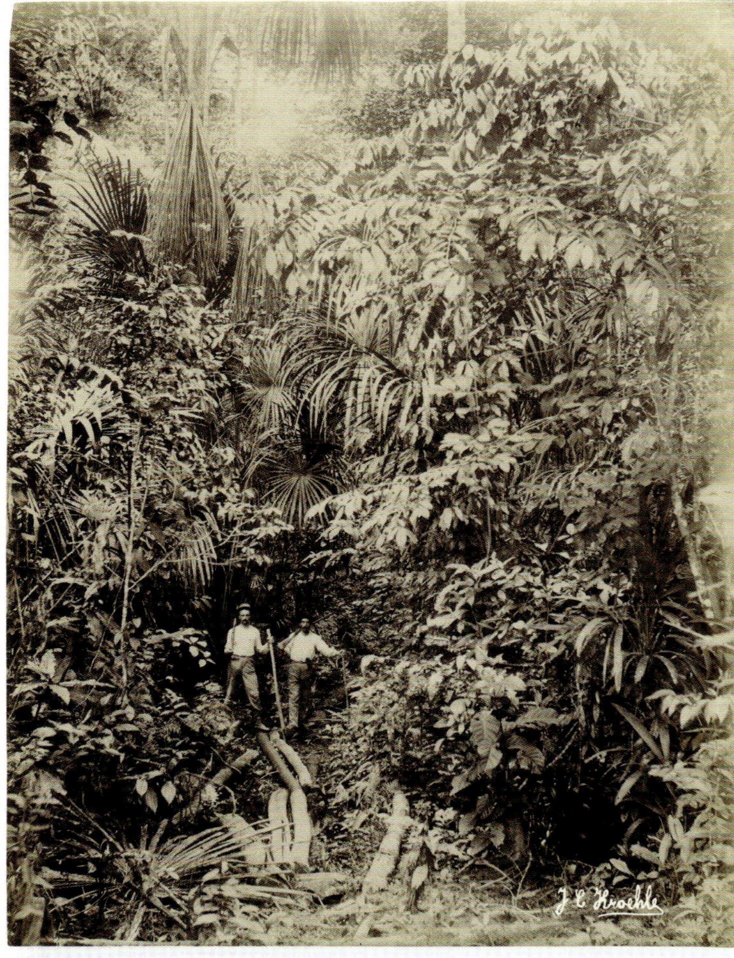

Charles Kroehle in the
Upper Amazon valley, Peru,
ca. 1890s

In summer 2016, Arguiñe Escandón sent Yann Gross, a Swiss photographer who often works in the Amazon, a postcard with a photograph by Charles Kroehle. It was one of many pictures the German photographer made while documenting Peru between 1888 and 1891. Escandón, a Spaniard and a photographer herself, added a friendly warning: "I hope you won't end up like him."

Although much is known about Georg Hübner, the German ethnographer and photographer with whom Kroehle traveled in eastern Peru, Kroehle's fate has been open to speculation— some say he disappeared after he was shot with a poisonous arrow in the rain forest. As Escandón and Gross considered their predecessors' earlier photographs, they saw an opportunity to make collaborative work. Almost a century and a half later, they traveled in Peru with the legend of Kroehle as a kind of anti-field guide.

While Hübner and Kroehle intended to produce a visual documentation of Indigenous Peoples and to send back prints for sale, Escandón and Gross were not as unquestioning of the implications of being foreign and the ties between power and representation. "We didn't want to bring back trophies, but tried to understand a bit more, even if the result was that we realized that we were totally ignorant," Gross said. "It was a good lesson in humility."

They traveled along the Pachitea, Ucayali, and Nanay Rivers, living among the Ashaninka, Shipibo-Conibo, and Cocama peoples, who must fight for the guarantee and acceleration of communal land titling, based on their rights of self-determination, and for alternative development plans that respect existing ecosystems. Escandón and Gross are rightly ambivalent about Kroehle and Hübner, who were complicit in more than one form of colonial exploitation, working with rubber barons, fur traders, and gold diggers.

The question for Peru, then as now, is how it might reckon with the pressures of global capitalism while addressing the fact that its resources are taken from Indigenous Peoples, whose claim to the territory is several thousand years old. The scale of the Peruvian Amazon—comprising 60 percent of the country, while occupied by only 5 percent of its population—makes parts of it prone to be allotted to companies engaged in mining, oil exploration, and hydroelectric megaprojects. "A concept is needed," Gross said, mindful of the impact of climate change and decreased biodiversity, and the worldviews of the peoples he and Escandón spent time with, "where you are part of an ecosystem and in balance. It's to be face to face with other elements and not above it—a concept of equality, more relational than hierarchical."

If, as the artists have noted, Gross's earlier photographs from the Amazon were documentary in nature and Escandón's were invested in psychology, their collaboration, *Tamamuri* (2018–ongoing), has produced a mix of both enthusiasms. The photographs they have returned with so far—whether a portrait or a detail of marshland, whether varnished with light silver or delicate blue—convey the intricacy and totality of an ecological surround.

Escandón and Gross are as foreign as their predecessors. Yet work of this kind, invested in sensation instead of a romantic representation of an unfamiliar culture, is an inward rather than outward exploration—an intrepid adventure that nevertheless rejects the logic of the explorer as discoverer. Their photographs mark a process of participation. Foremost on their minds was the possibility that they could find a through line connecting self and environment, image and history.

Emmanuel Iduma, a critic and novelist, is the author, most recently, of *A Stranger's Pose* (2018).

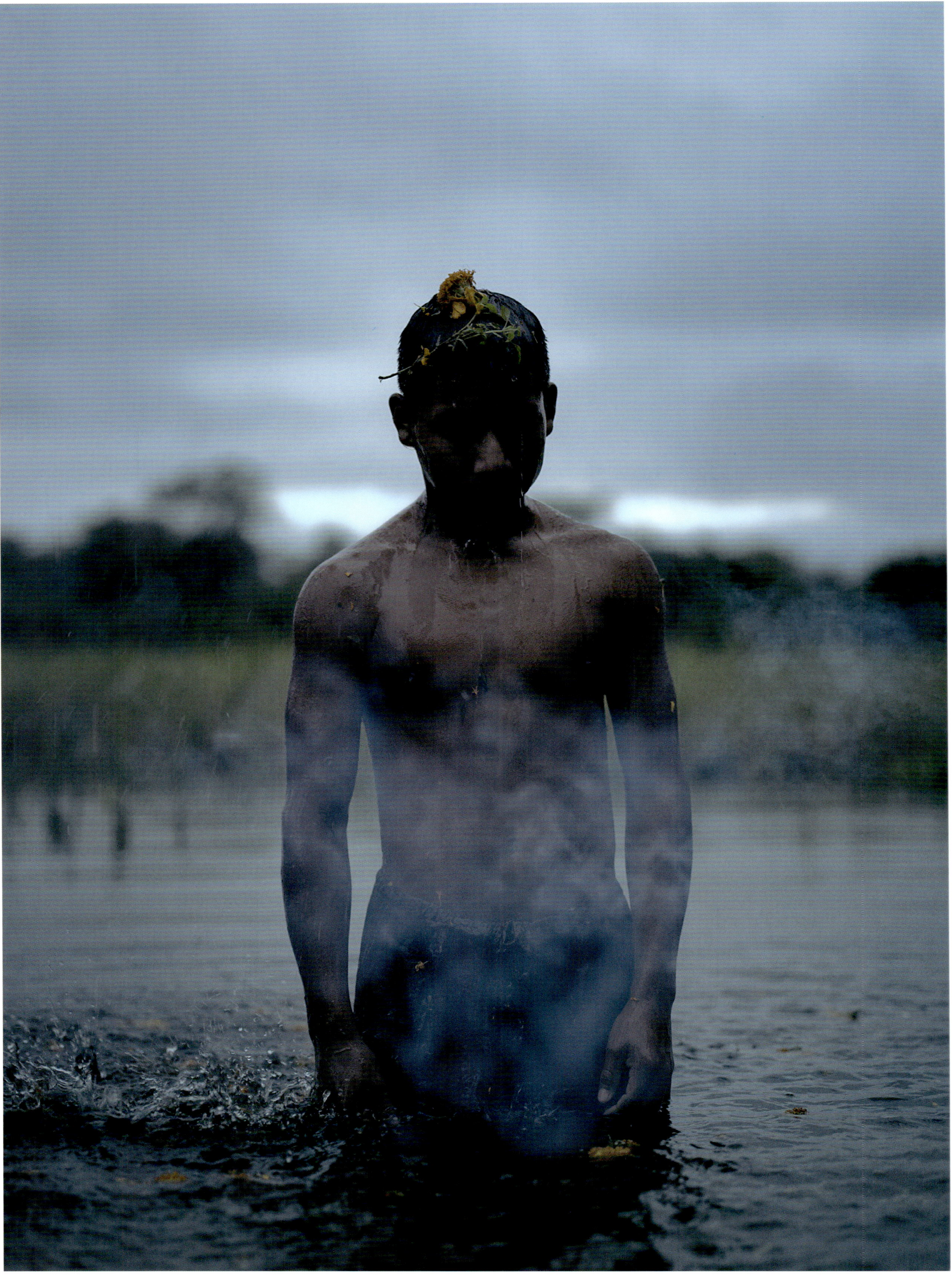

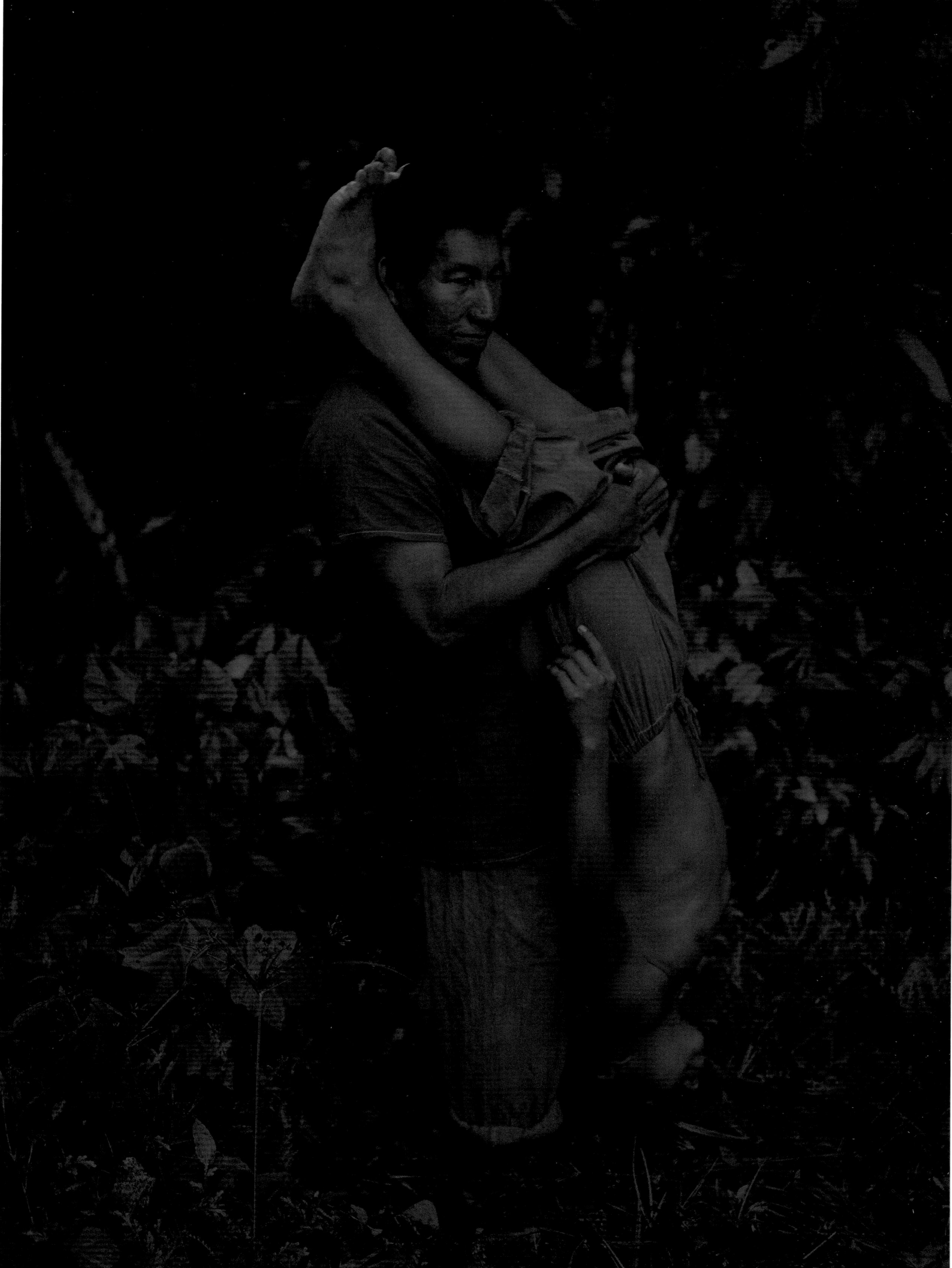

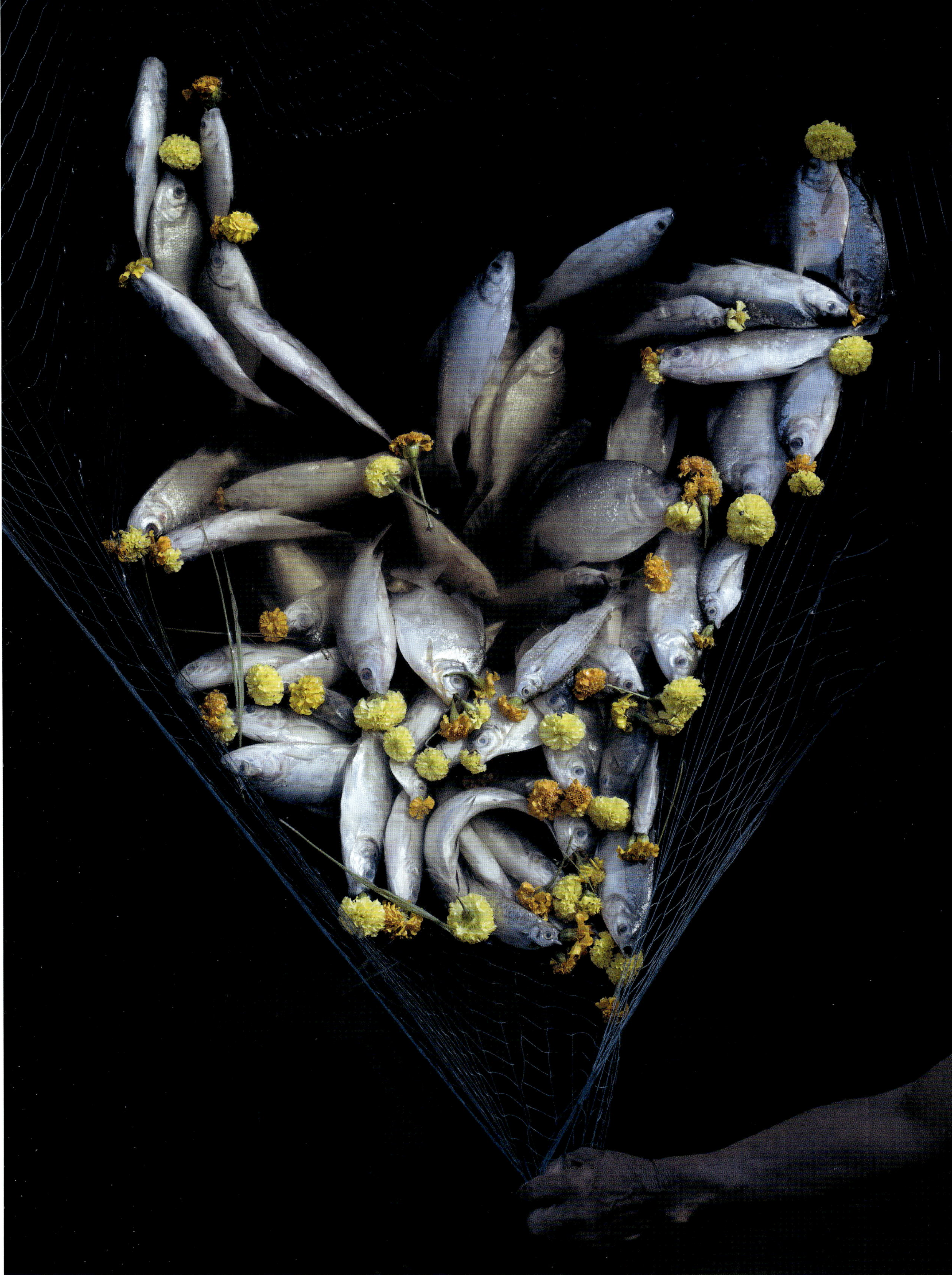

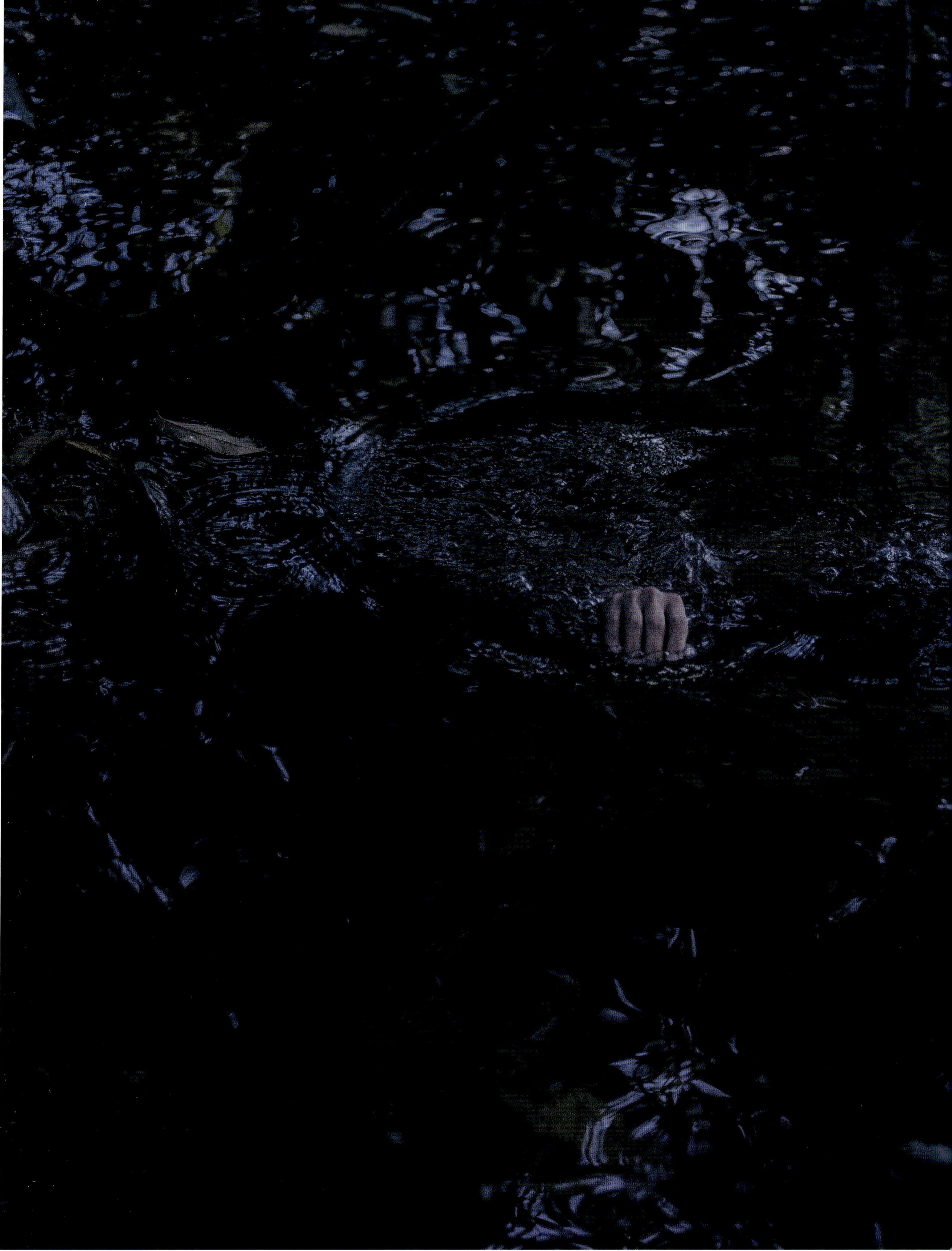

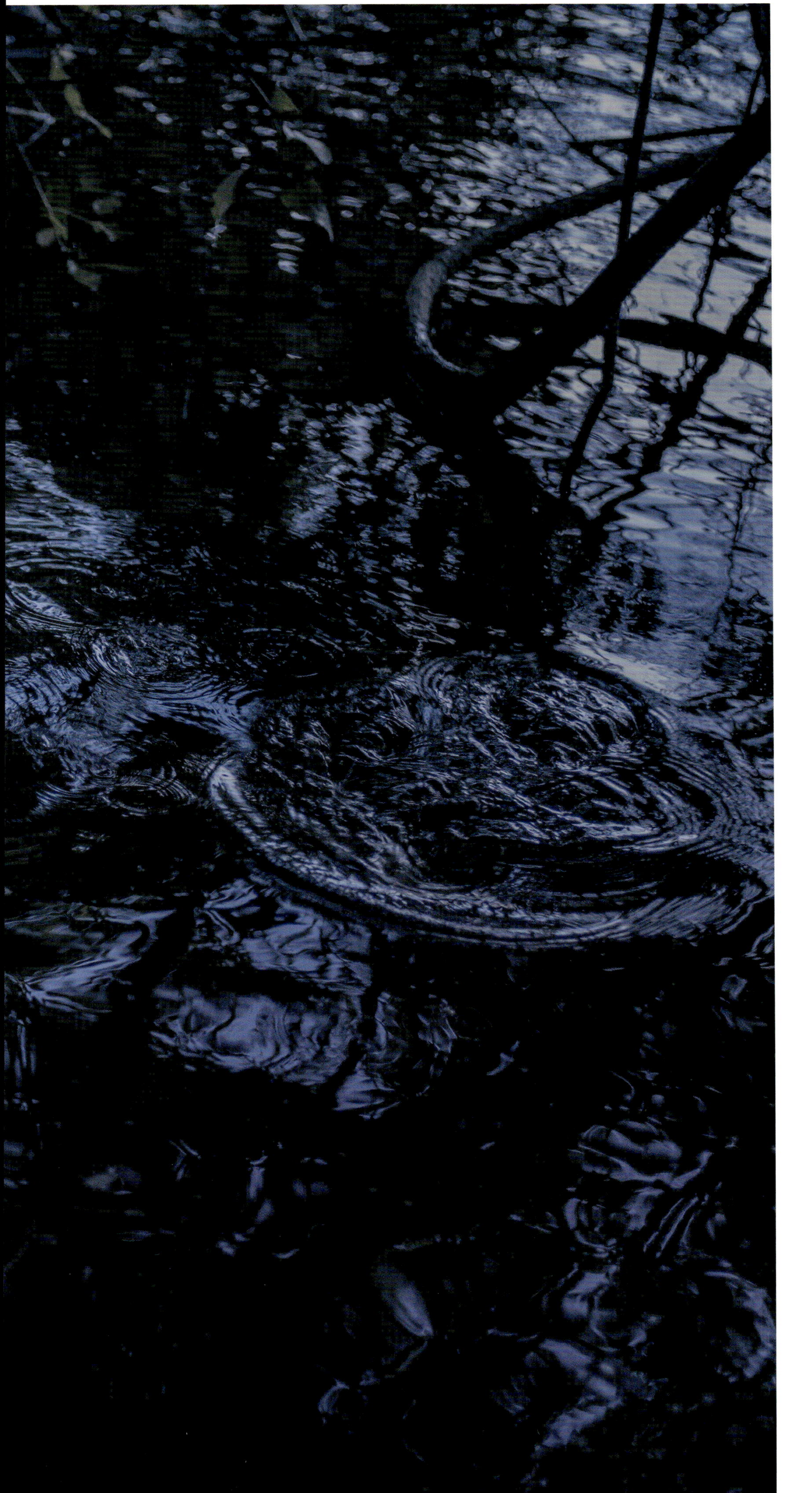

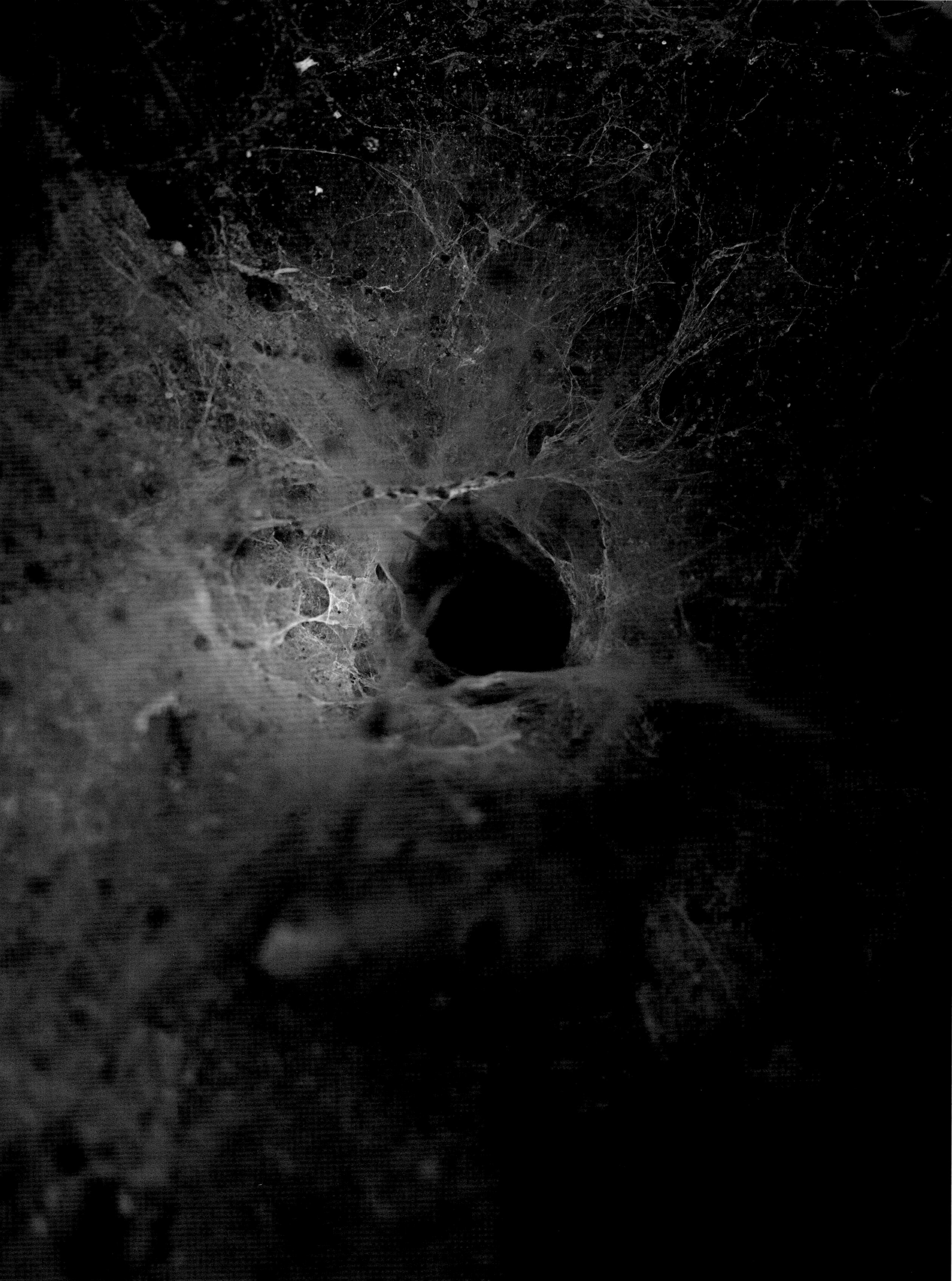

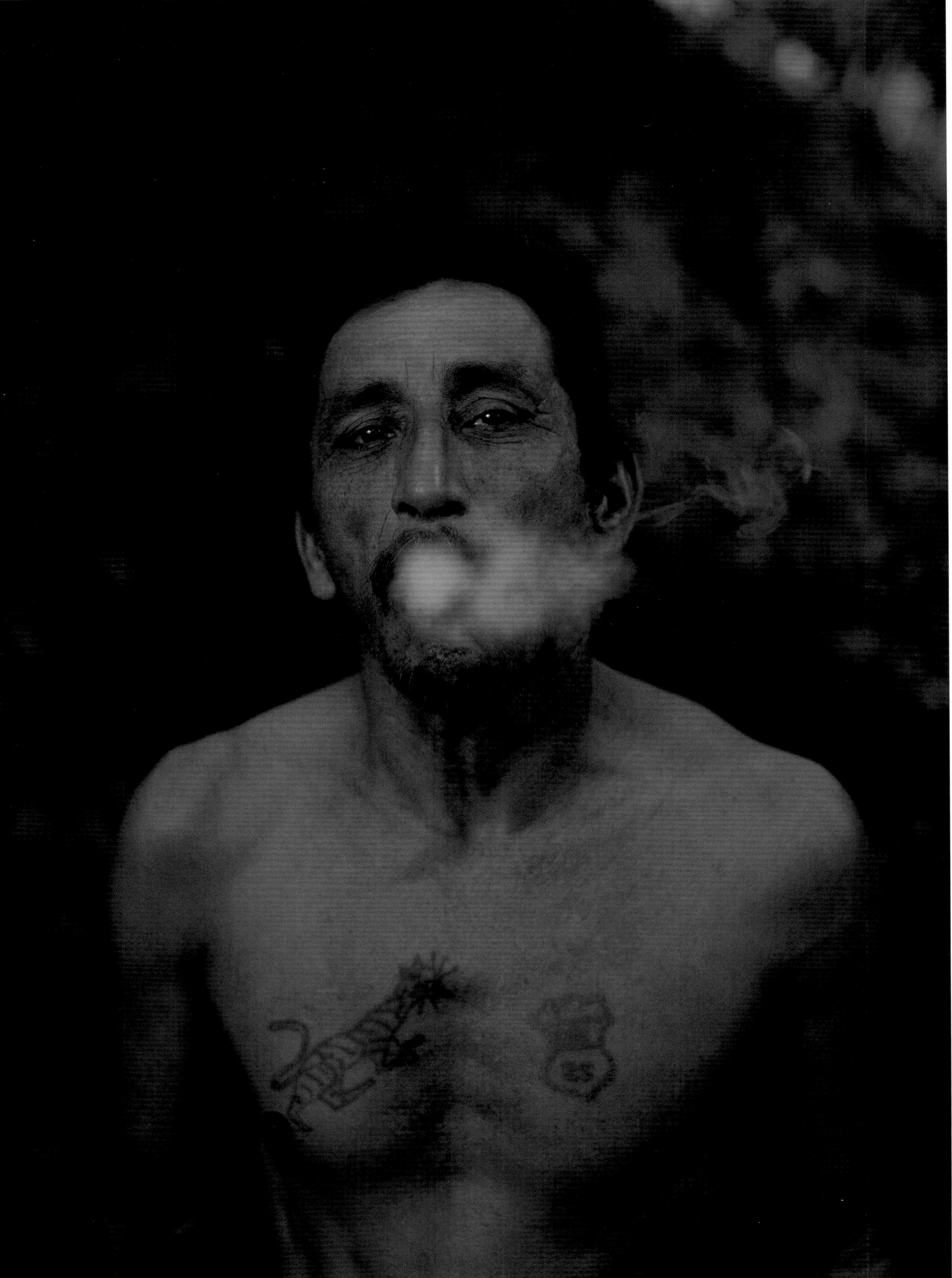

On October 19, 2018, to mark the Indian festival of Dussehra, the prime minister of India ceremonially shot an arrow into a giant effigy of Ravana—the enemy of Rama, the eponymous hero of the ancient Indian epic the *Ramayana*. Ravana's effigy was installed in Delhi, where thousands had gathered to watch this mythic spectacle of the destruction of evil. By some pyrotechnical sleight of hand, the prime minister's arrow, although missing its target, officially set off a blaze of firecrackers concealed inside Ravana's body that could not have had a salutary effect on the capital's already lethal air-quality index. Never mind if some of the nation's environmentally concerned citizens have been fighting a vain battle to get the burning of festive firecrackers banned by the country's highest court.

Like the endlessly retold battle between good and evil, the coming together and falling apart of history and myth, past and present, fact and fiction, politics and faith simultaneously define how an epic like the *Ramayana* is inextricable from the bewilderingly varied fabric of everyday life in contemporary India. I could not help recalling the ritual encounter between Ravana and the prime minister when sitting down a few weeks after Dussehra to think about Vasantha Yogananthan's *A Myth of Two Souls*, his ongoing, seven-part, photographic retelling of the *Ramayana*, a project that began in 2013. The fourth chapter, *Dandaka*, is set in the Dandaka Forest—a place of punishment and exile for Rama; his newly married wife, Sita; and his devoted brother Lakshmana. At once a mythological topos and an actual topography, the modern-day Dandaka Forest region is spread across several Indian states, each with its own dynamic of environmental depredation and activism that is inseparable from political conflict, social change, economic development, and cultural diversity.

Yogananthan—who lives in Paris, whose mother is French, and whose father is from Sri Lanka—travels repeatedly through-out India to make this body of work. There is no guarantee that he will find any easily indexical correspondence between the places he photographs and those he reads about in the innumerable versions of the *Ramayana*, even if their names happen to be the same. He allows, therefore, the timelessness, the historicity, as well as the contemporaneity of the epic to be refracted through the actual and the immediate in a series of encounters with what Louis Malle calls "the real" in his seven-part documentary *L'Inde fantôme: Reflexions sur un voyage* (1969). Yogananthan colors these refractions, literally and metaphorically, by employing a number of narrative and pictorial traditions, both archaic and modern, and collaborating with a variety of local storytellers, artists, and actors.

The dreamy forestland in these hand-painted photographs—in which objects, gestures, expressions, and interactions turn mysteriously, even sinisterly, allusive—happens to be situated in the actual jungles of Odisha, Chhattisgarh, and Karnataka. These are states where daily encounters with migration, displacement, industrialization, and environmental damage are lived out within, and between, communities that are defined by shifting configurations of class and caste, as well as tribal, linguistic, religious, and political identities, allegiances, and interests. This is a postcolonial history that stretches back from the present toward the strategically credulous invention of a mythical past, which may be invoked to validate the oppressive violence of a nation-state when it chooses to play its games of power, corruption, exploitation, and greed in the guise of a modern and secular democracy.

The most memorable poetic expression of such an experience of history—at once living, lived, and imagined—might be found in the essay "Three Hundred *Ramayanas*: Five Examples and Three Thoughts on Translation," written and revised between 1985 and 1991, by the late A. K. Ramanujan, the multilingual poet, translator, and scholar. In this essay, Ramanujan includes his translation of the description of a river that begins a retelling of the epic *Irāmāvatāram* by Kampan, the medieval poet: "Turning forest into slope, / field into wilderness, / seashore into fertile land, / changing boundaries, / exchanging landscapes, / the reckless waters / roared on like the pasts / that hurry close on the heels / of lives. / Born of Himalayan stone / and mingling with the seas, / it spreads, ceaselessly various, / one and many at once." For Yogananthan, Ramanujan's essay informed his understanding of how the *Ramayana* continues to generate its proliferating afterlives in the modern world. But Ramanujan's essay ran into trouble in India—with those who bigotedly prefer to enshrine the One as opposed to celebrating the Many.

Vasantha Yogananthan

Aveek Sen

All photographs from
A Myth of Two Souls,
chapter 4, *Dandaka*, 2018
Courtesy the artist

Aveek Sen is a writer based in Kolkata.

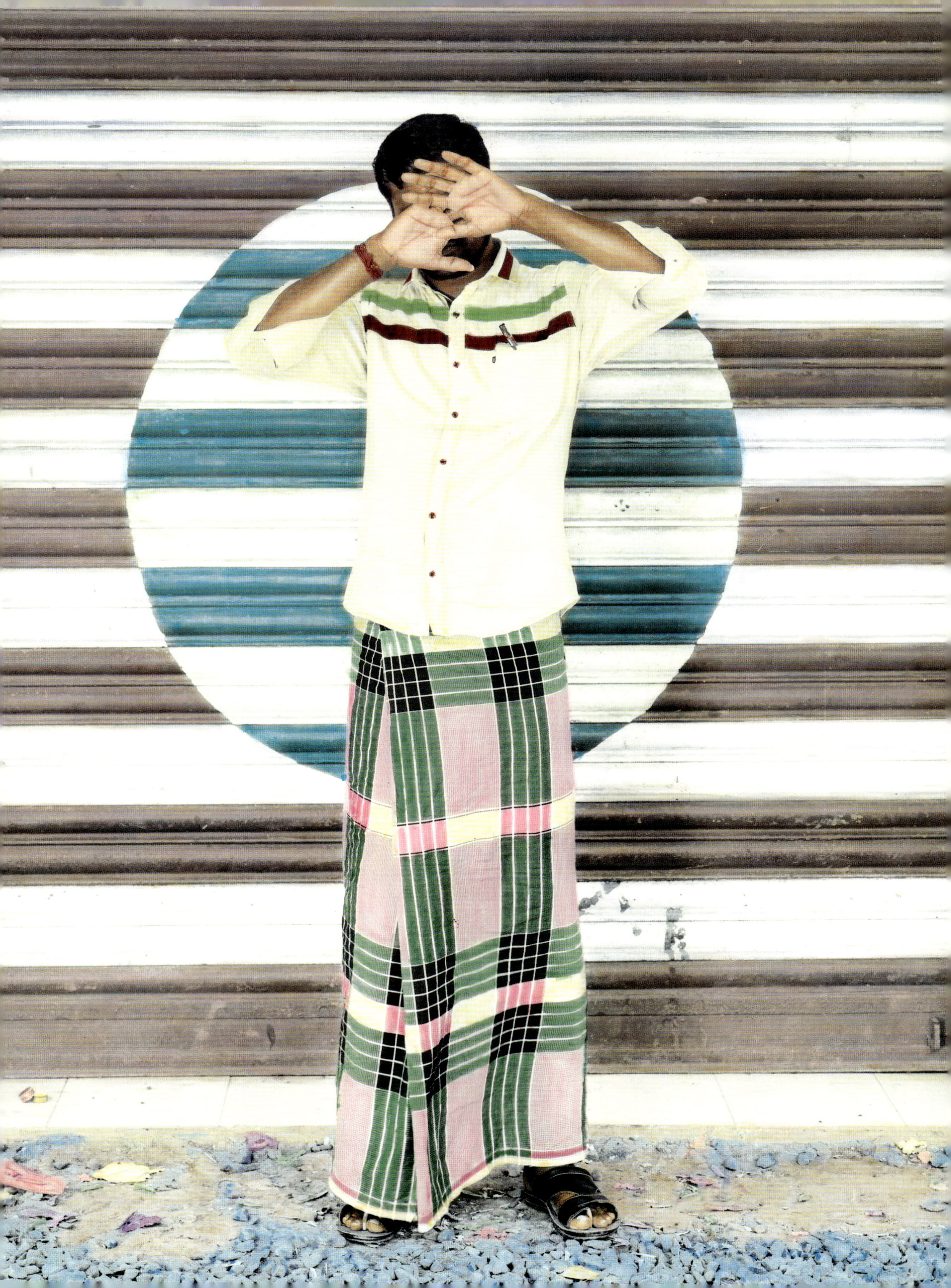

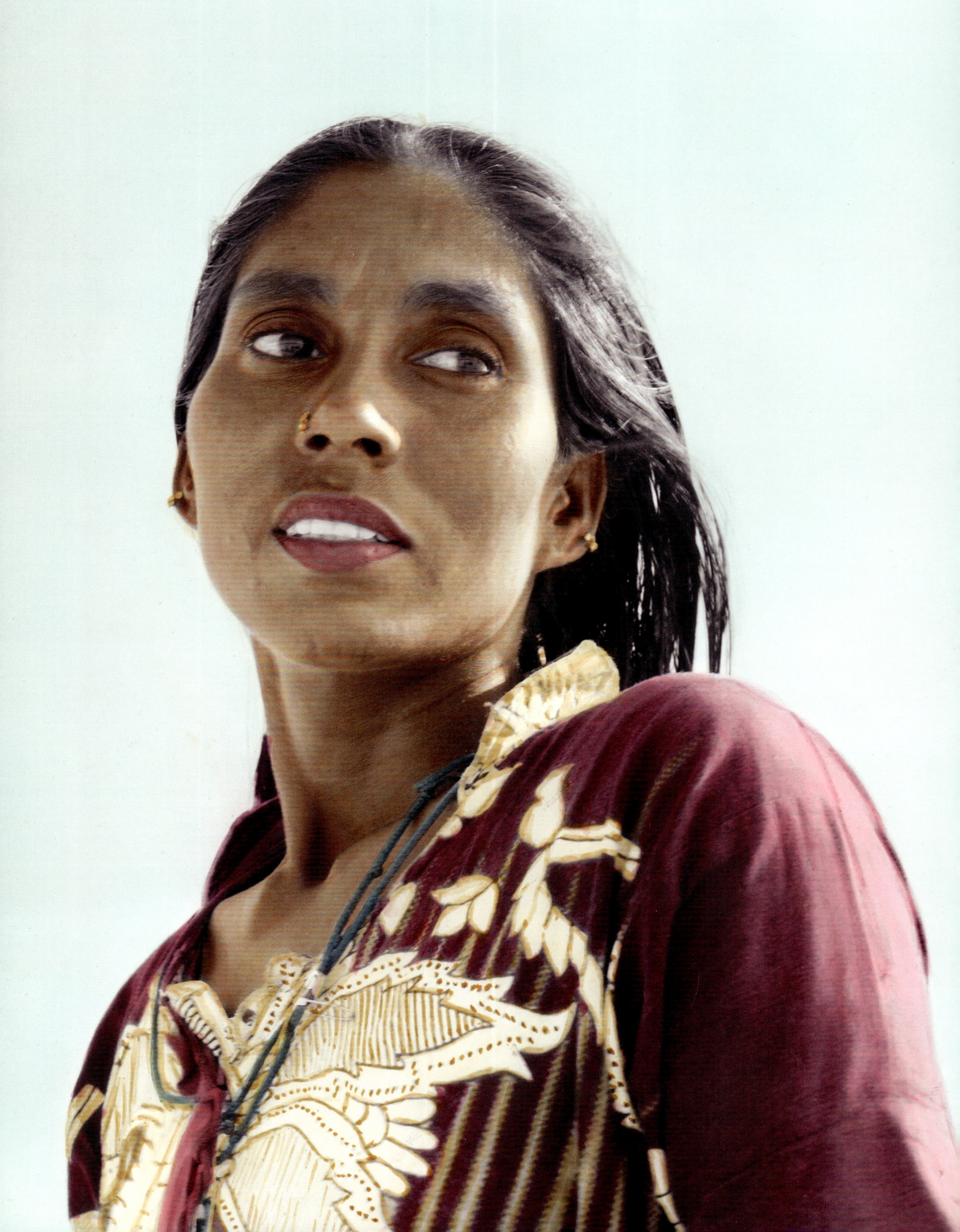

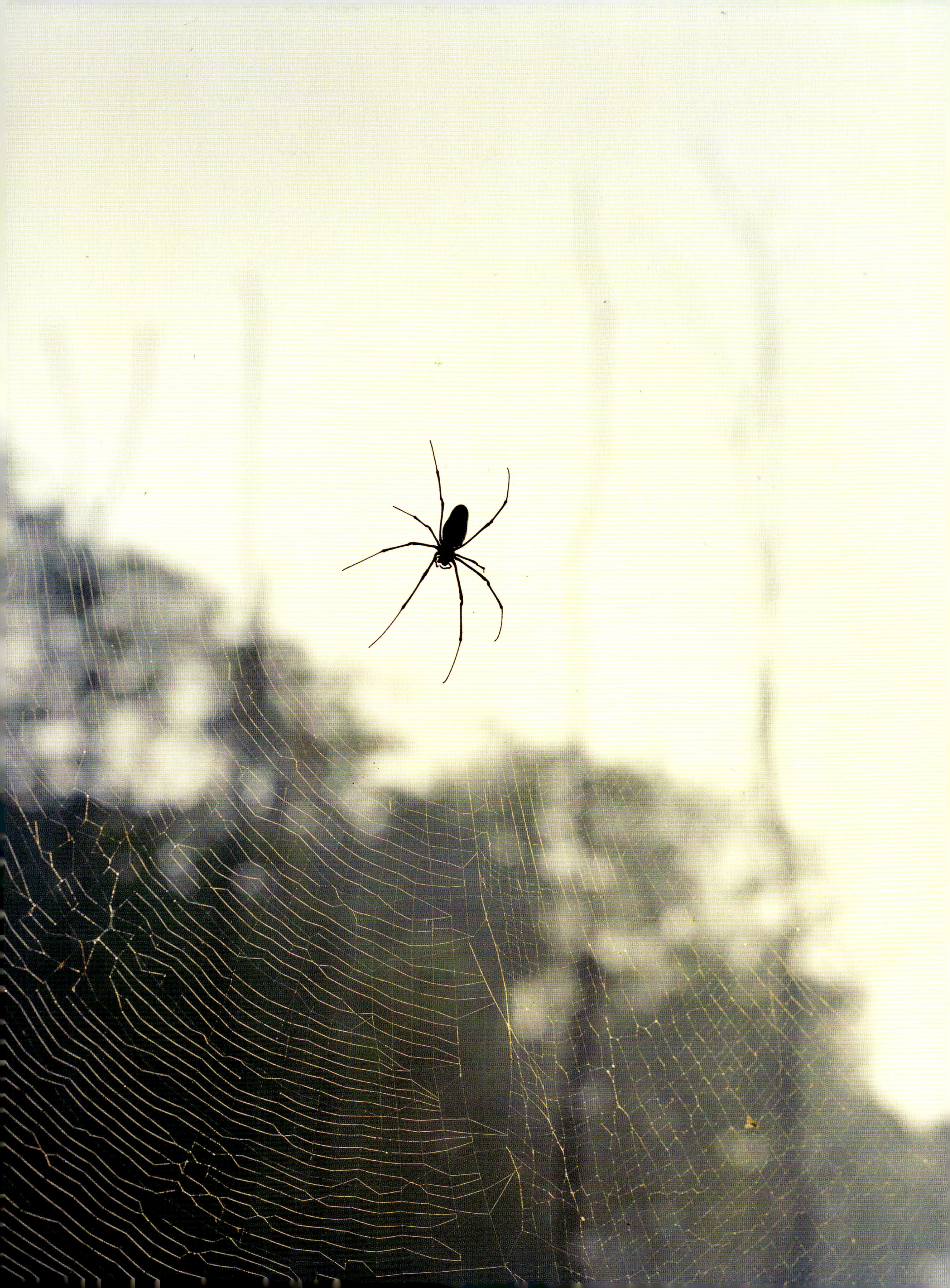

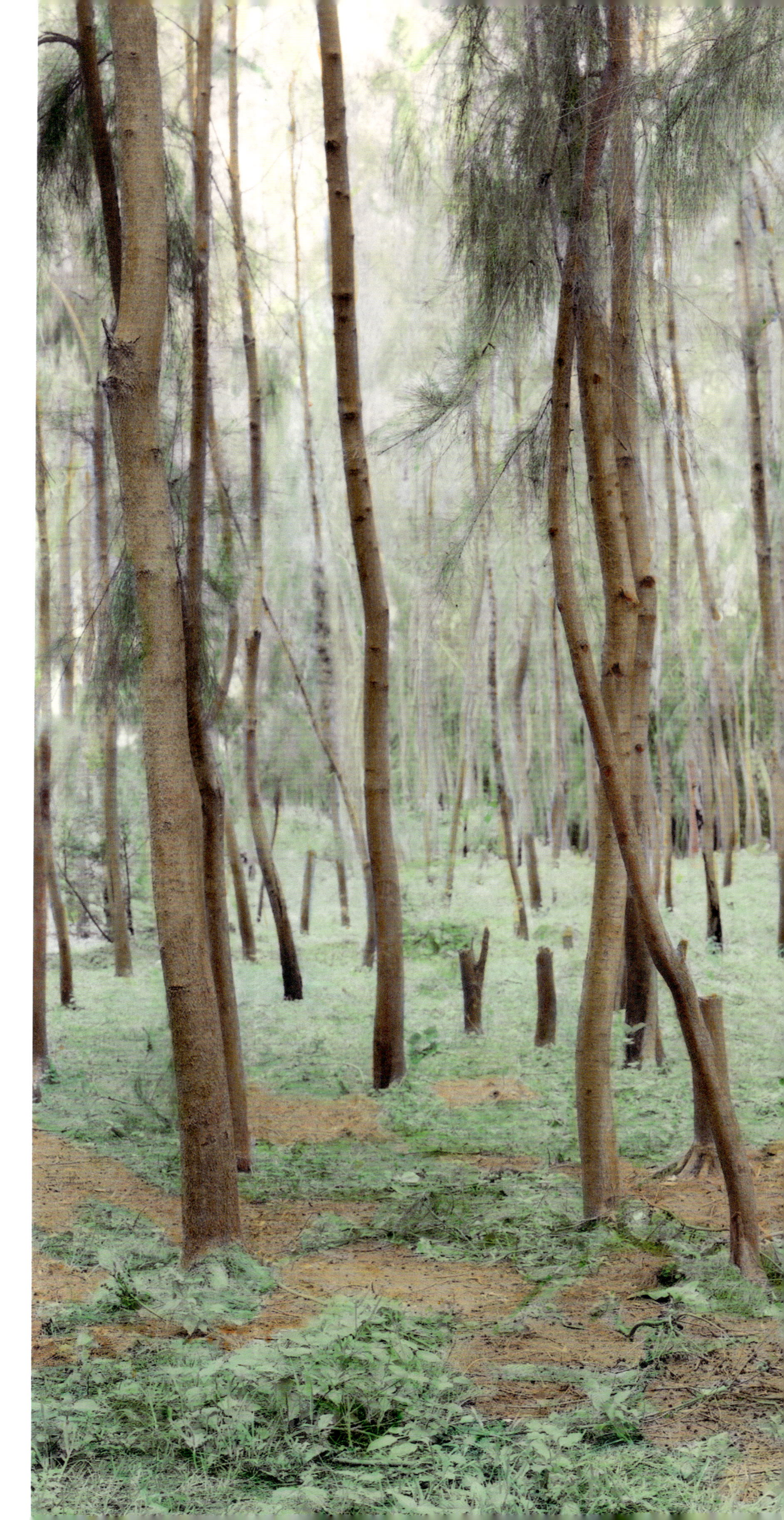

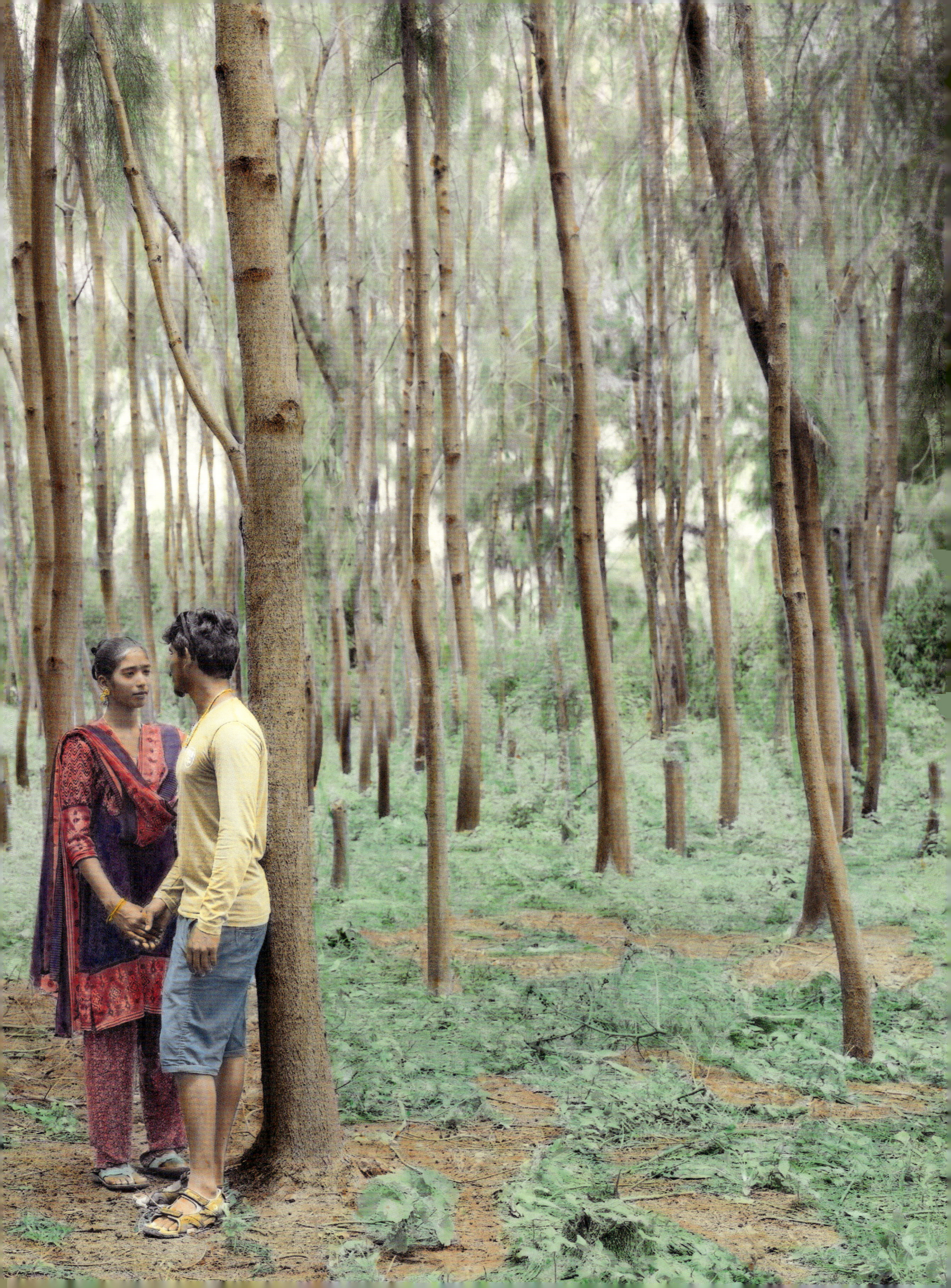

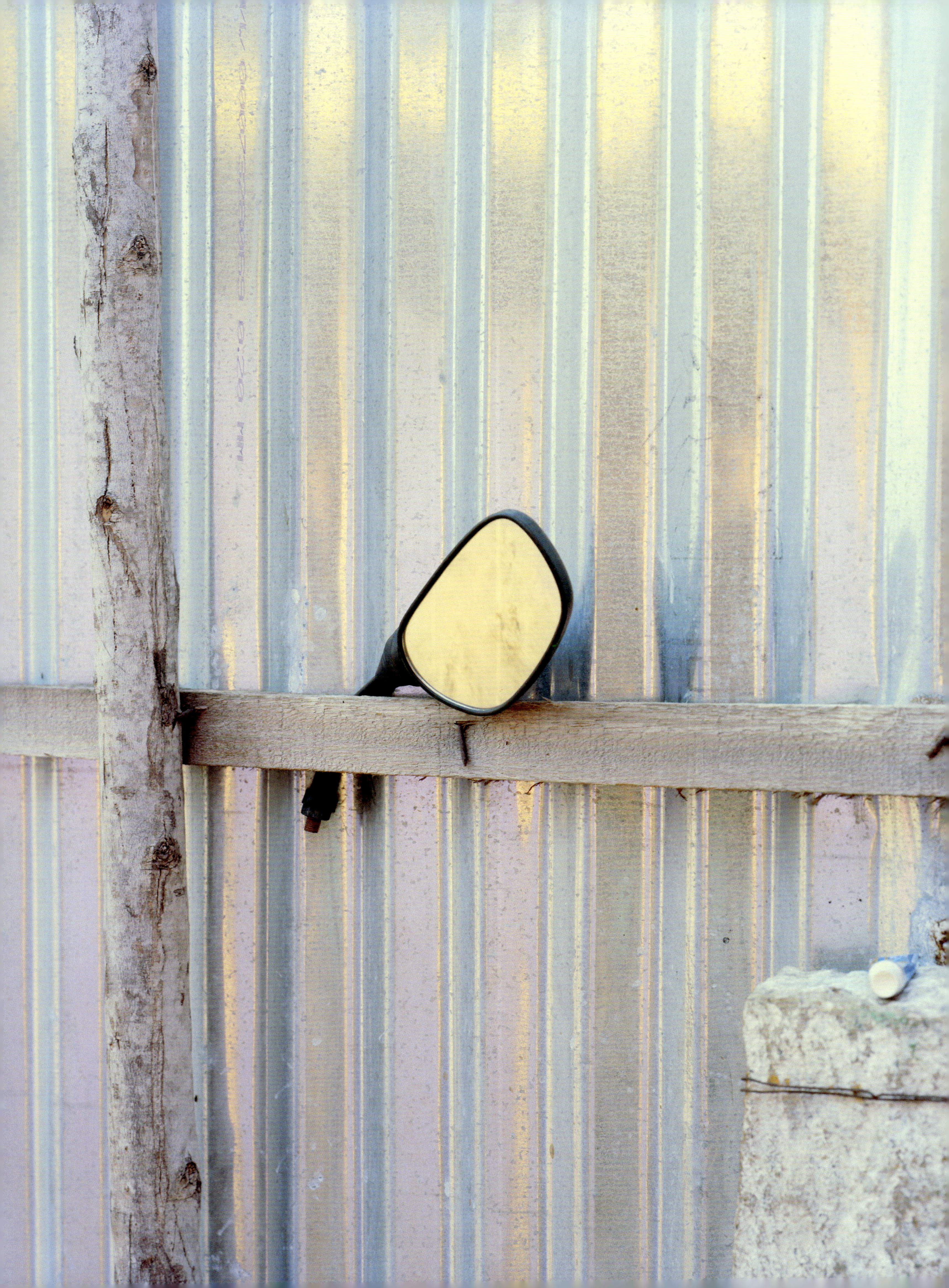

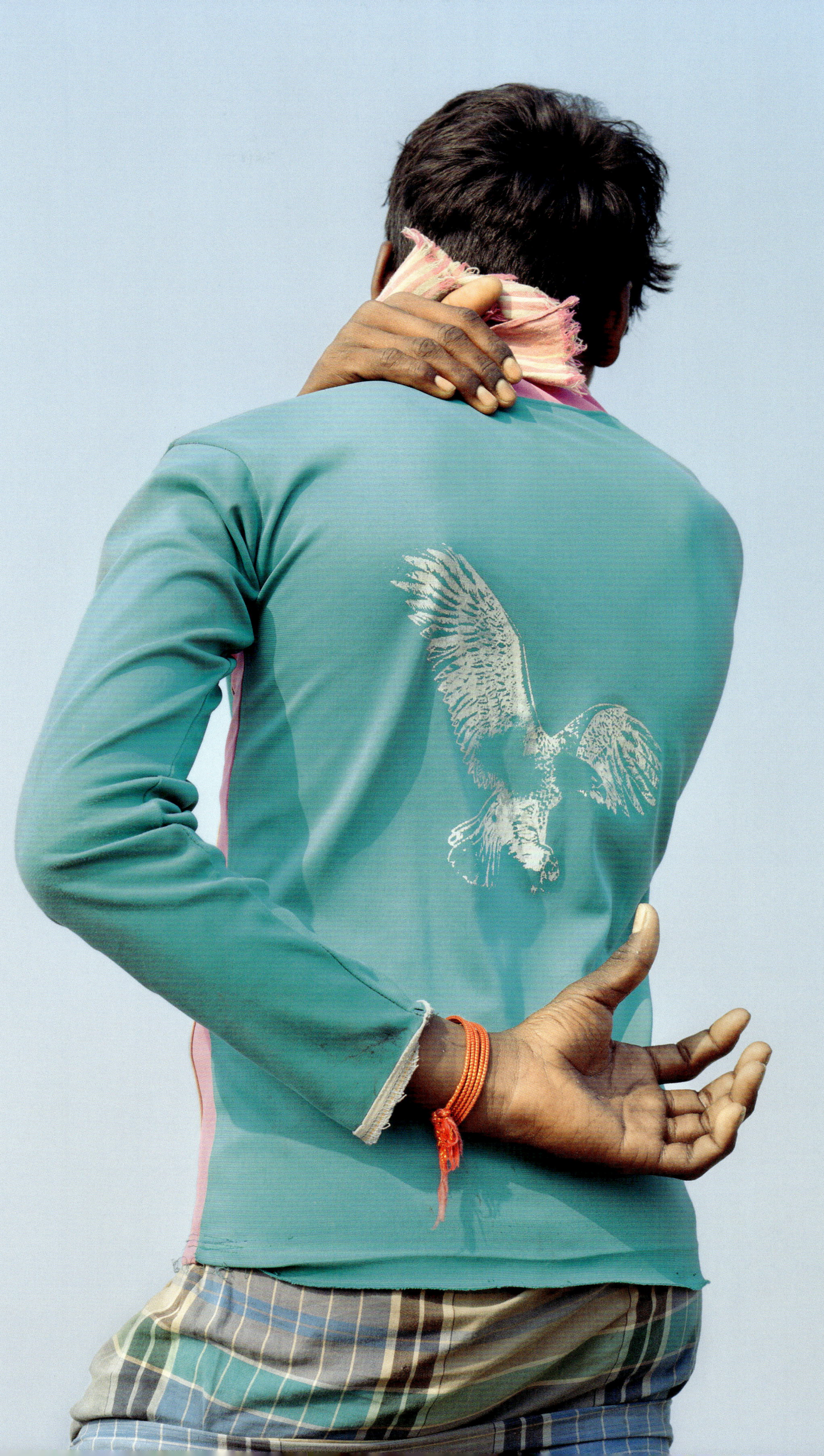

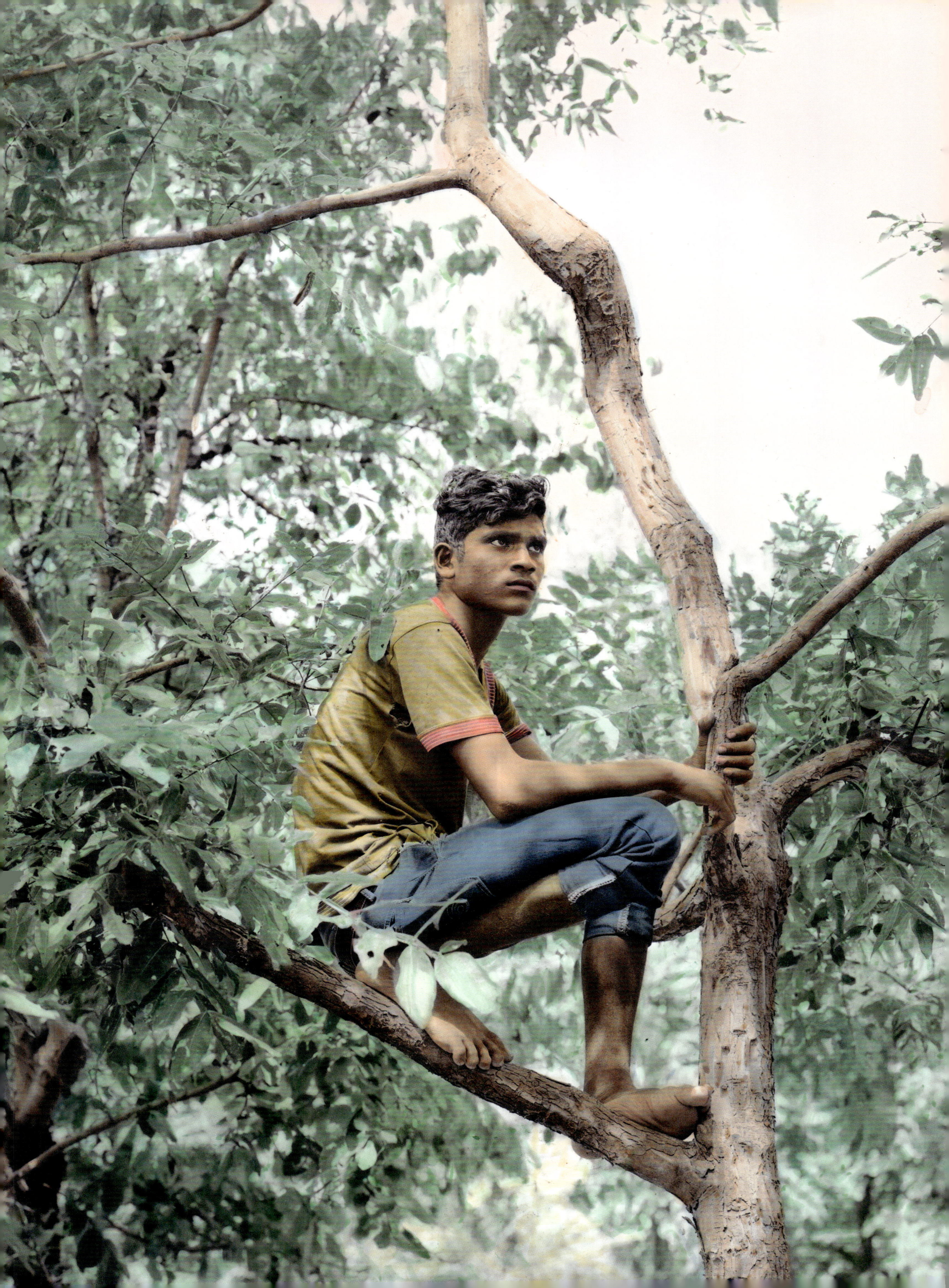

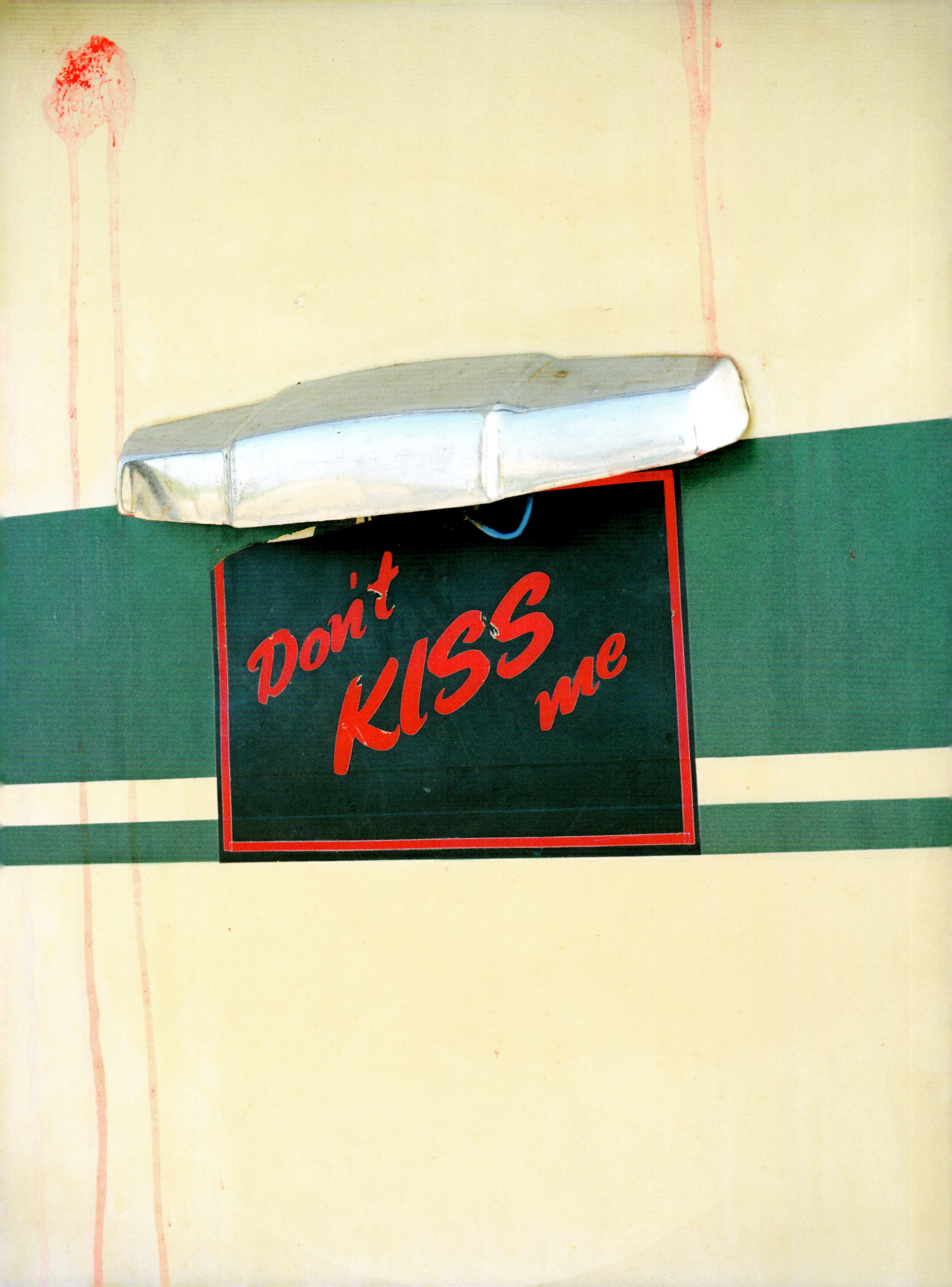

Don't KISS me

David Benjamin Sherry

Bill McKibben

Before you've seen the West, you've seen the West—landscape photographs of the region, especially those by Ansel Adams, are so deep in our nation's collective imagination that you have to work to actually see Half Dome, in California, or Shiprock, in New Mexico, even when you're standing there with your hiking boots on.

David Benjamin Sherry's recent pictures help us see again. Sherry is known for his fascination with color, for his analog techniques, and for what some have called his "queer revision" of the rugged and macho legacy of western landscape photography. His images of several national monuments, photographed last year, carry the same level of detail as Adams's iconic pictures, the sublime clarity of the haze-free western summer afternoon. But drenched in unexpected and unreal color, they get you to take a second look.

And in this case, a second look is helpful for any number of reasons.

For one, looking backward, the great protected areas of the nation are not simply blank slates, empty wastes. They were often the homelands of this continent's original inhabitants, and so they tell, among other things, the stories of our nation's original shame. Their very emptiness is a reminder of what we did—all the more telling when the petroglyphs left behind at places like Bears Ears, the national monument in Utah, make

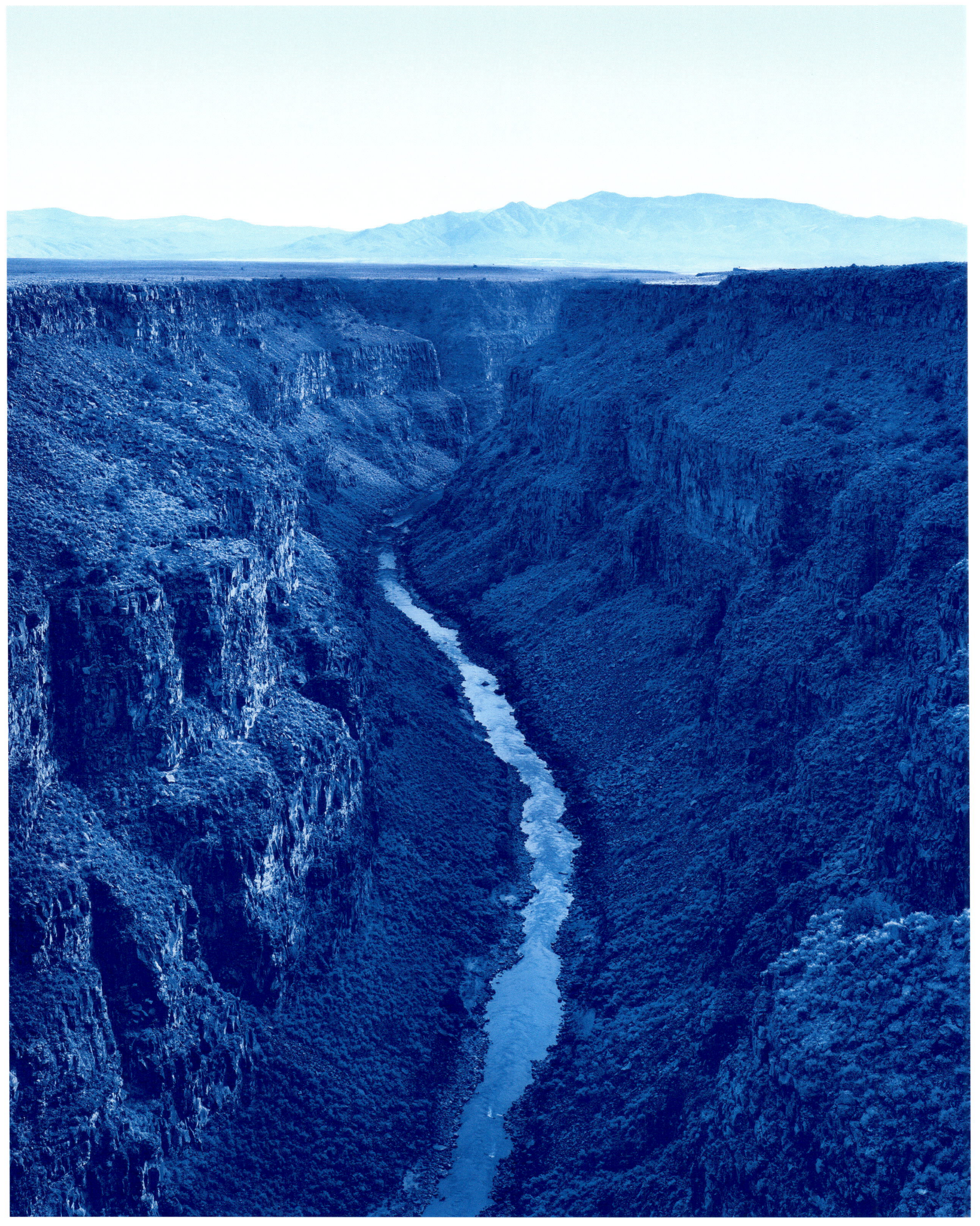

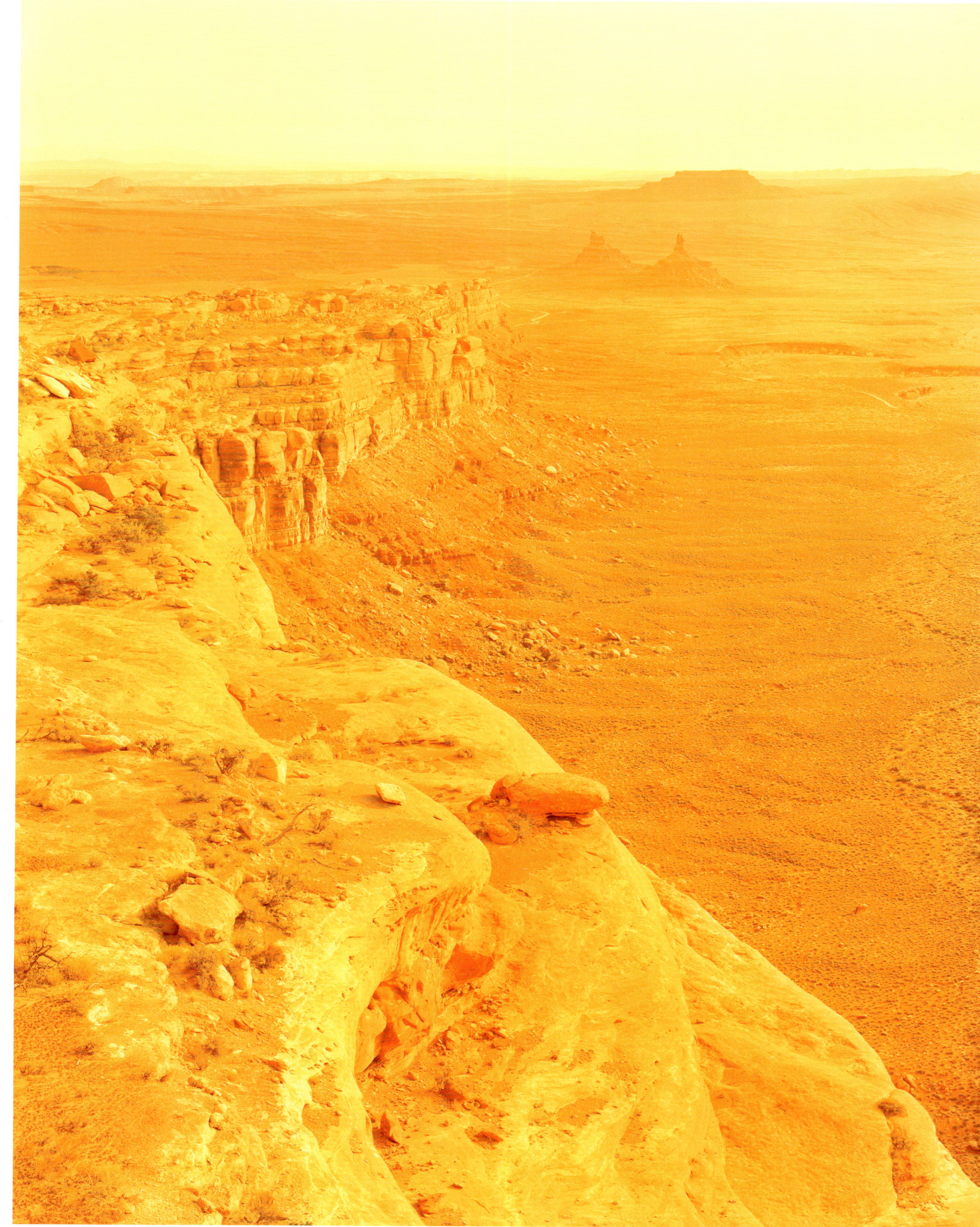

Muley Point I, Bears Ears
National Monument, Utah

clear what a bustling place it once was. These lands are as sacred to Indigenous cultures as they ever were, but there's a tragic quality to that reverence now.

For another, looking forward, these same lands are no longer as sacred to the colonizing tradition as they once were. One of the great boasts of its legacy was the protected landscape: in the late nineteenth century and throughout the twentieth, we felt ourselves rich enough to methodically put aside large tracts of land for the benefit of the rest of creation, or the future, or our idea that there was something lovely about wildernesses, even ones we might not see. Congress never got more poetic than with the Wilderness Act of 1964, with its commitment to protecting places "where the earth and its community of life are untrammeled by man, where man himself is a visitor who does not remain." Aside from the questions already raised about who was there originally, and aside from the obnoxious use of *man* that belies the text's birthdate, the statute still marks something powerful: even in the middle of America's great postwar boom, the understanding that we needed something more than we had.

But we don't think that anymore. Or at least, at the moment, those in charge don't think that. President Donald Trump, among his endless provocations, has begun trying to roll back the protections of an earlier era, beginning with the national monuments pictured in Sherry's images. For no reason other than to undo the work of the bigger souls who came before him, the petulant boy king has begun to take apart the network of protected areas that is one of the country's great legacies. Actually, of course, there is another reason: the fossil fuel industry covets these lands, just as it covets the Arctic, and the offshore lease holdings along the North American coasts, and pretty much every other piece of real estate on the continent. Not content with merely destroying the planet's climate, it must also do what it can to wreck the loveliness that has been set aside.

Somehow the saturated and unsettling colors of Sherry's photographs of Bears Ears and Grand Staircase-Escalante National Monument, in Utah, and the Río Grande del Norte National Monument, in New Mexico, among other western vistas, help us see all that splendor, all that history, and all those politics more clearly, or at least glimpse that something has gone wrong and is now going wronger in these places that have long been a comforting part of the landscape of the mind. No longer retreats or redoubts from the overwhelming bleat of our wired world, they are contested places. We must fight to make sense of them, and we must fight to preserve them, and we must fight to make sure that in their preservation they connect us back to the people who wandered them originally.

Iconic images have their place—but iconoclasm has its place too.

Bill McKibben, a writer and environmentalist, is the author, most recently, of *Falter: Has the Human Game Begun to Play Itself Out?* (2019).

Muley Point II, Bears Ears
National Monument, Utah

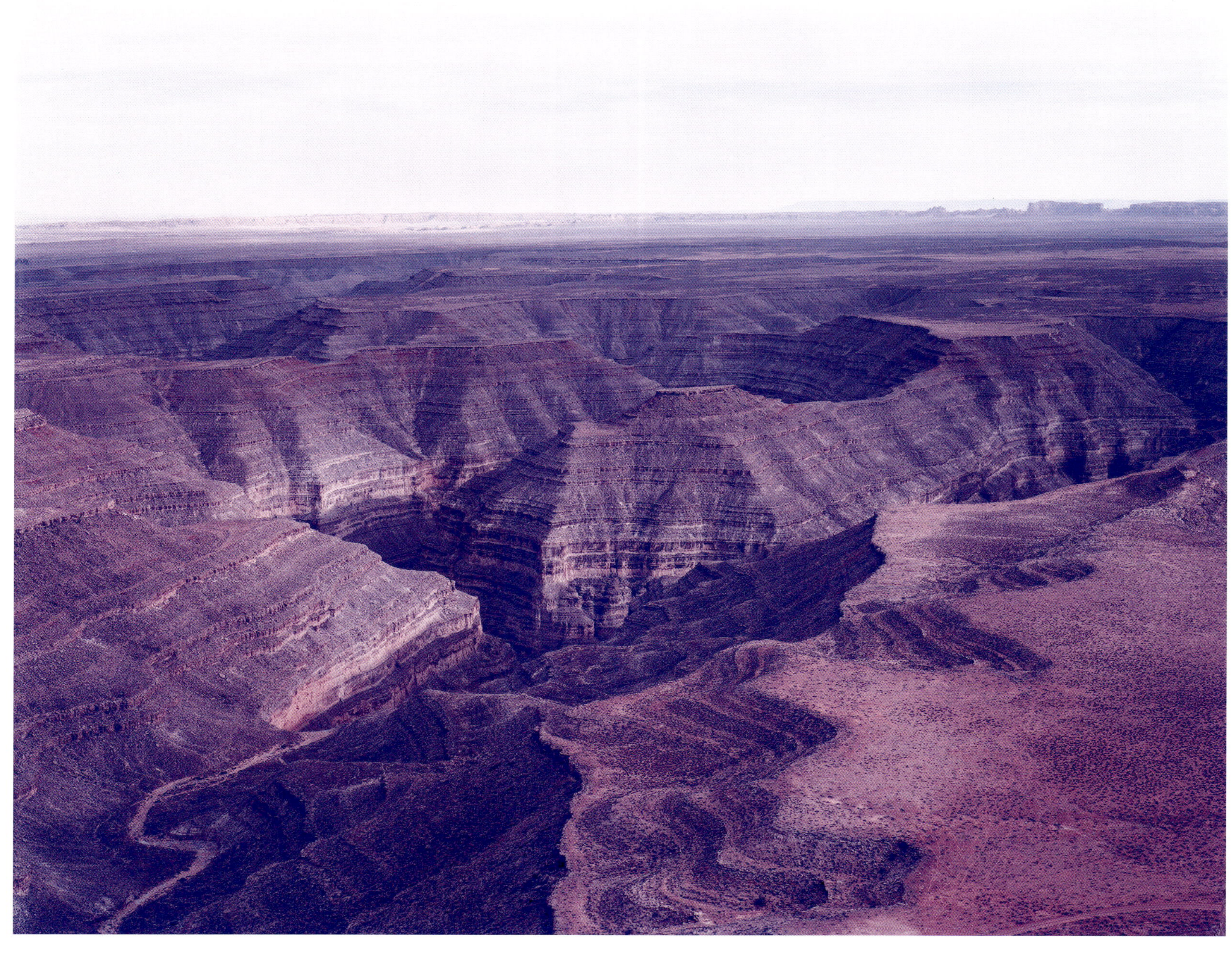

Sotol cactus, Organ
Mountains-Desert Peaks
National Monument,
New Mexico

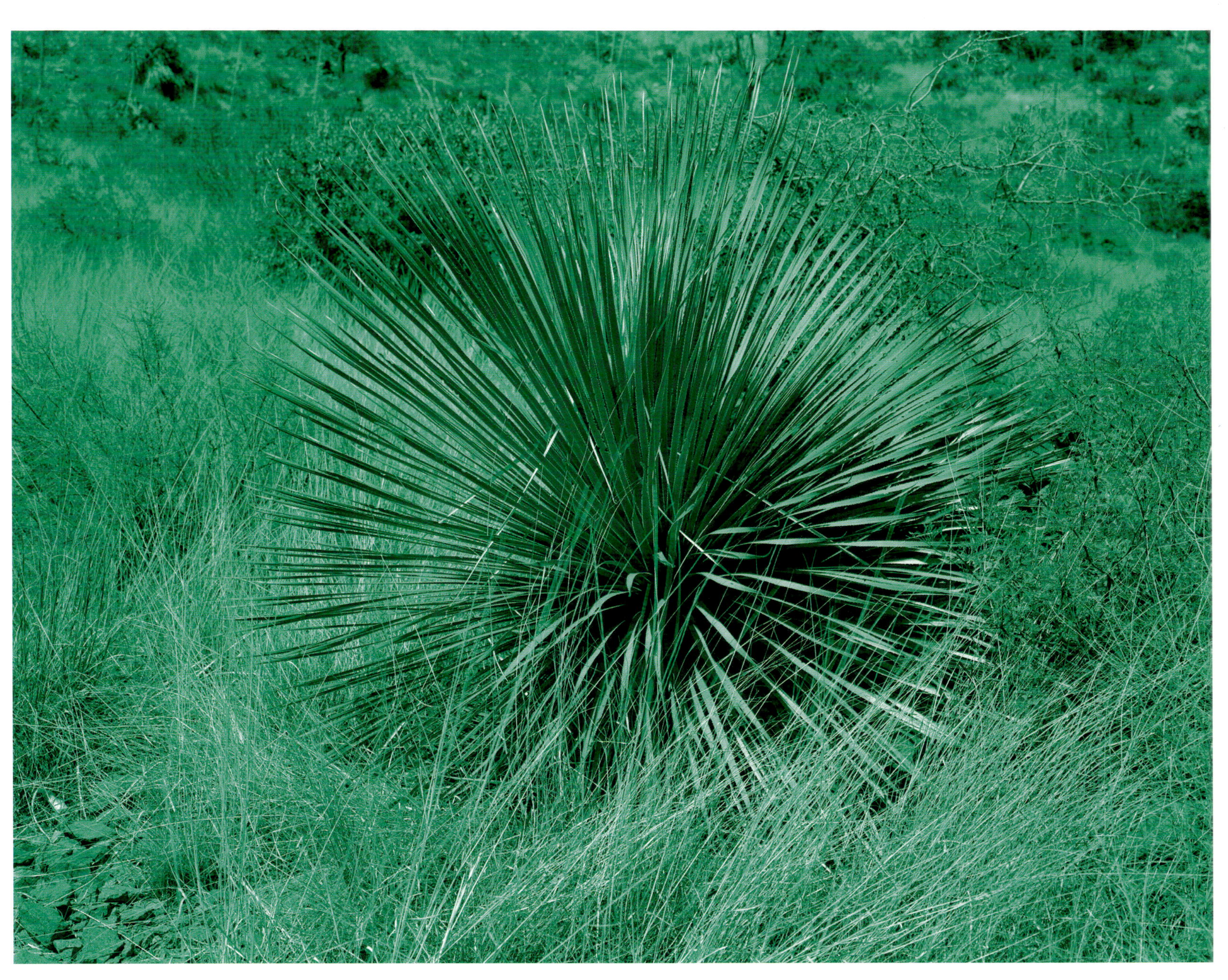

Cottonwood tree, Bears
Ears National Monument,
Utah

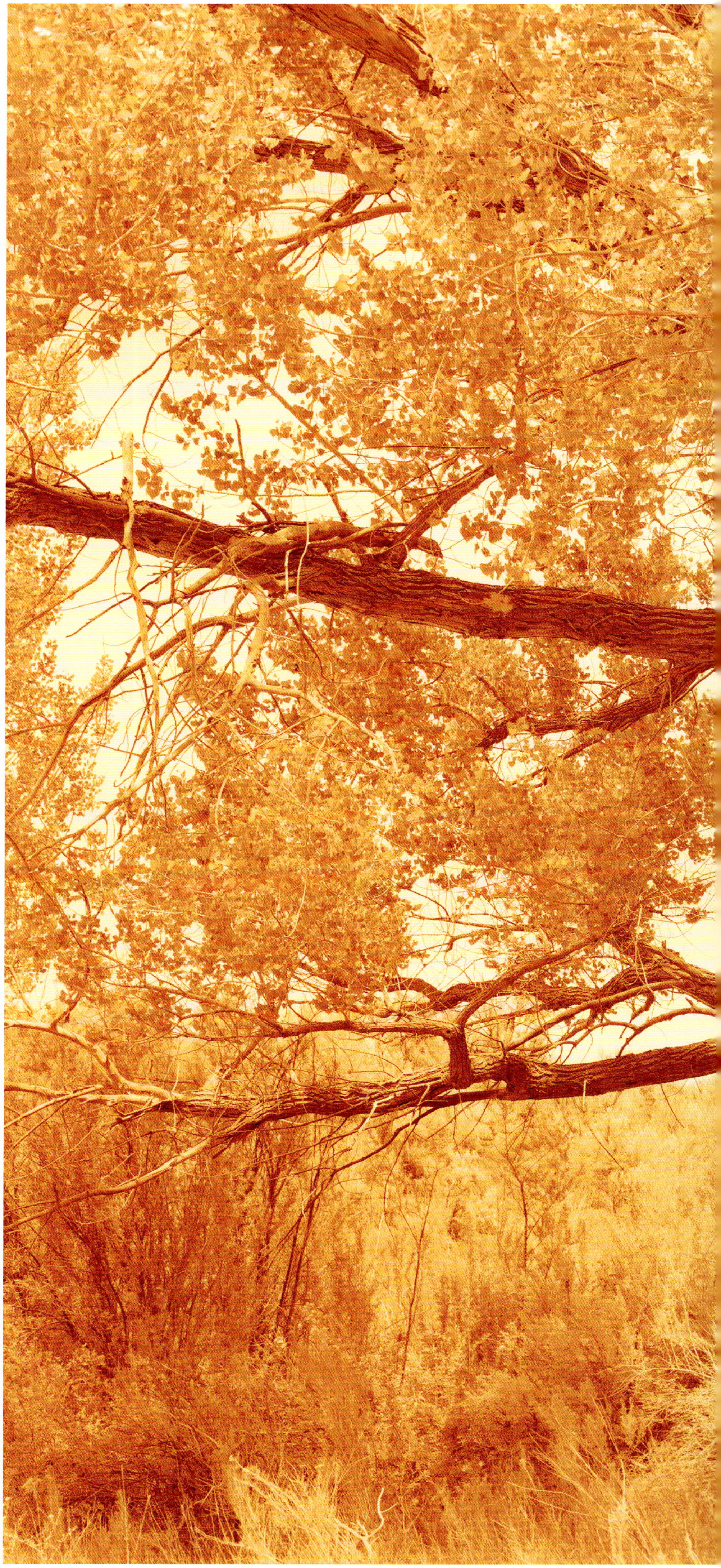

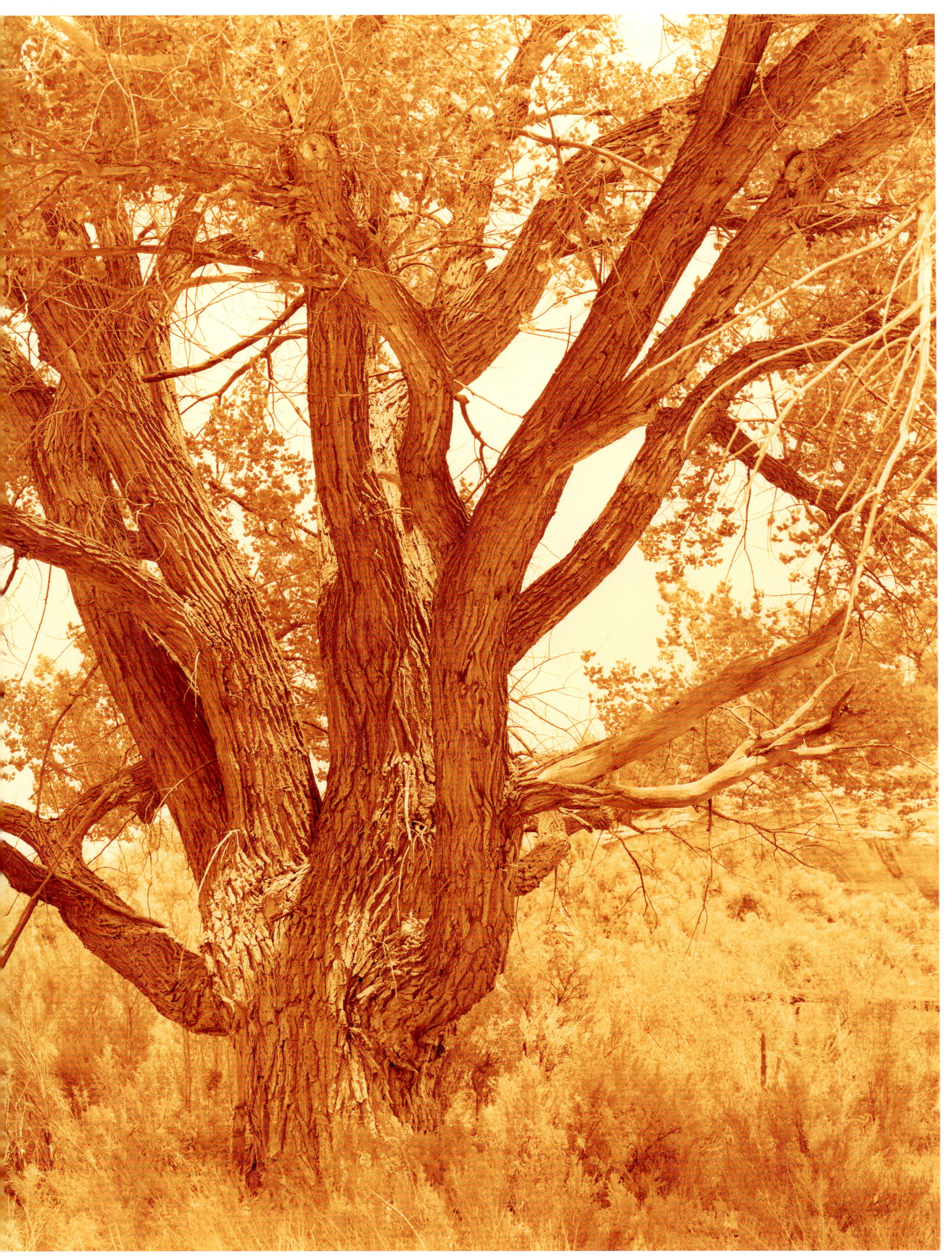

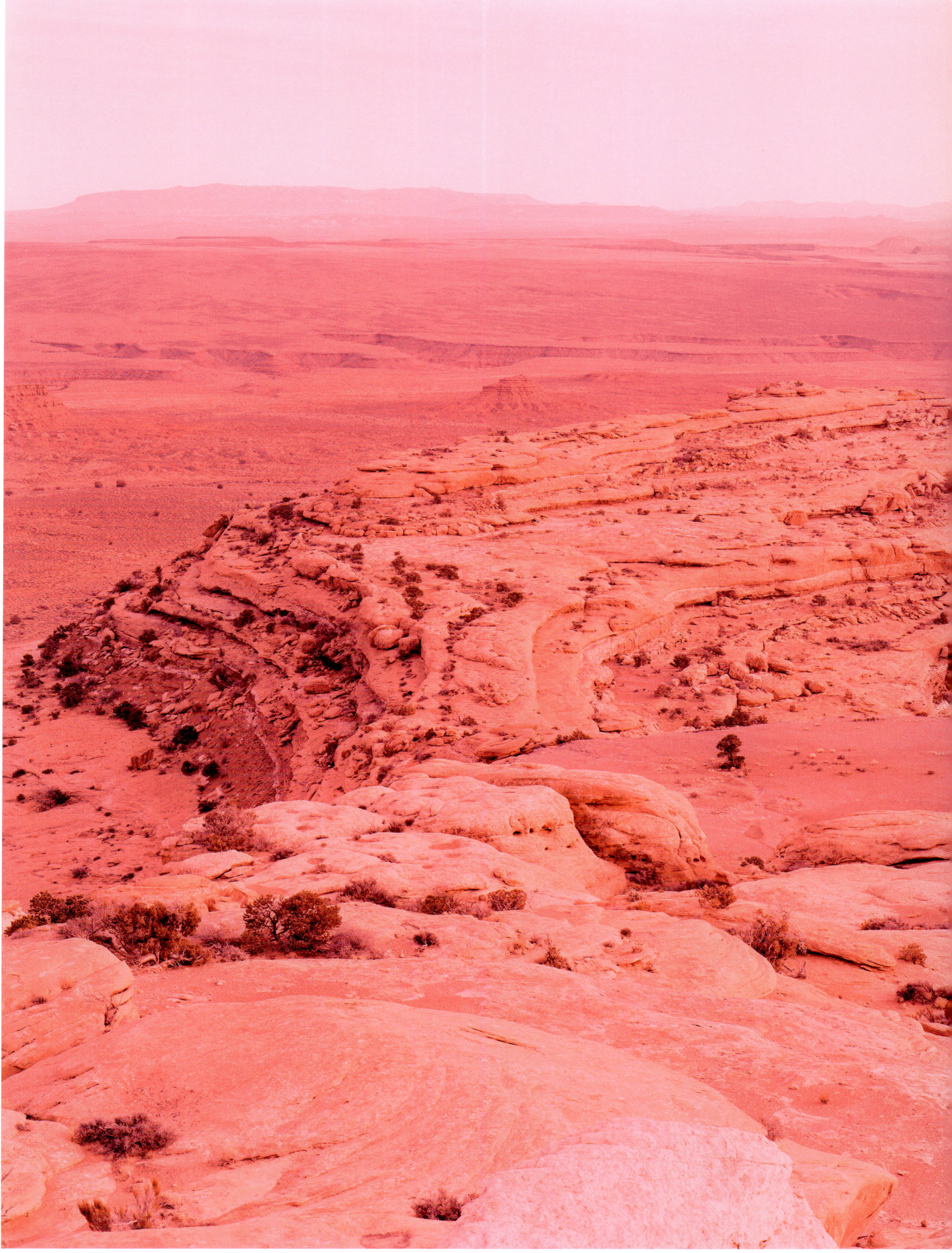

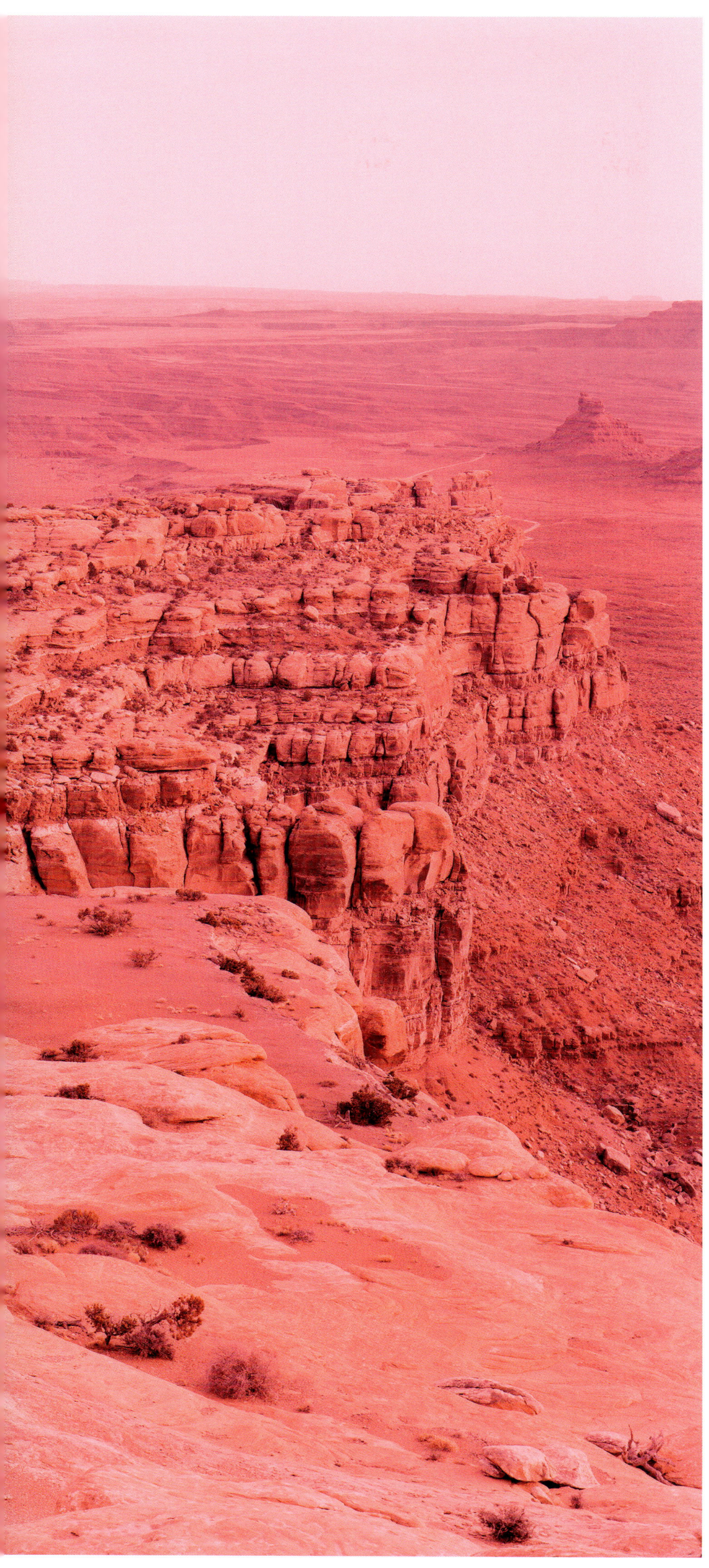

**Looking toward Valley
of the Gods, Bears Ears
National Monument, Utah**

All photographs 2018
Courtesy the artist and
Salon 94

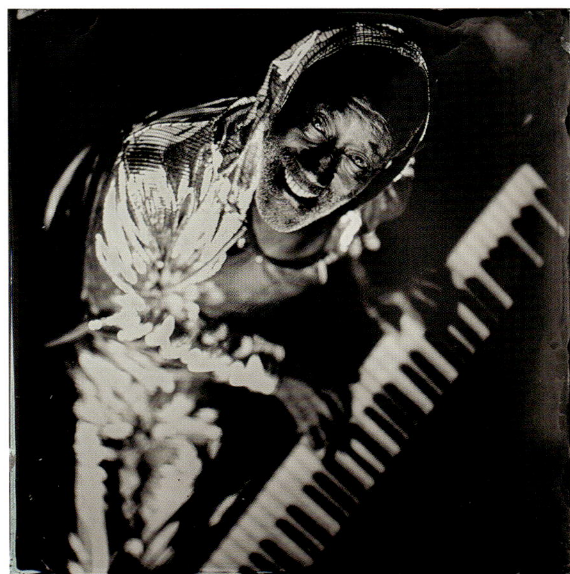

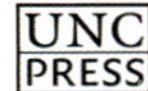

Fotomuseum Winterthur

Anne Collier –
Photographic

23.02.–26.05.2019

Sophie Calle
Un certain regard

08.06.–25.08.2019

SITUATIONS/
Photo Text Data

23.02.–02.06.2019

M|C|A PHOTOGRAPHY

www.mica.edu/aperture

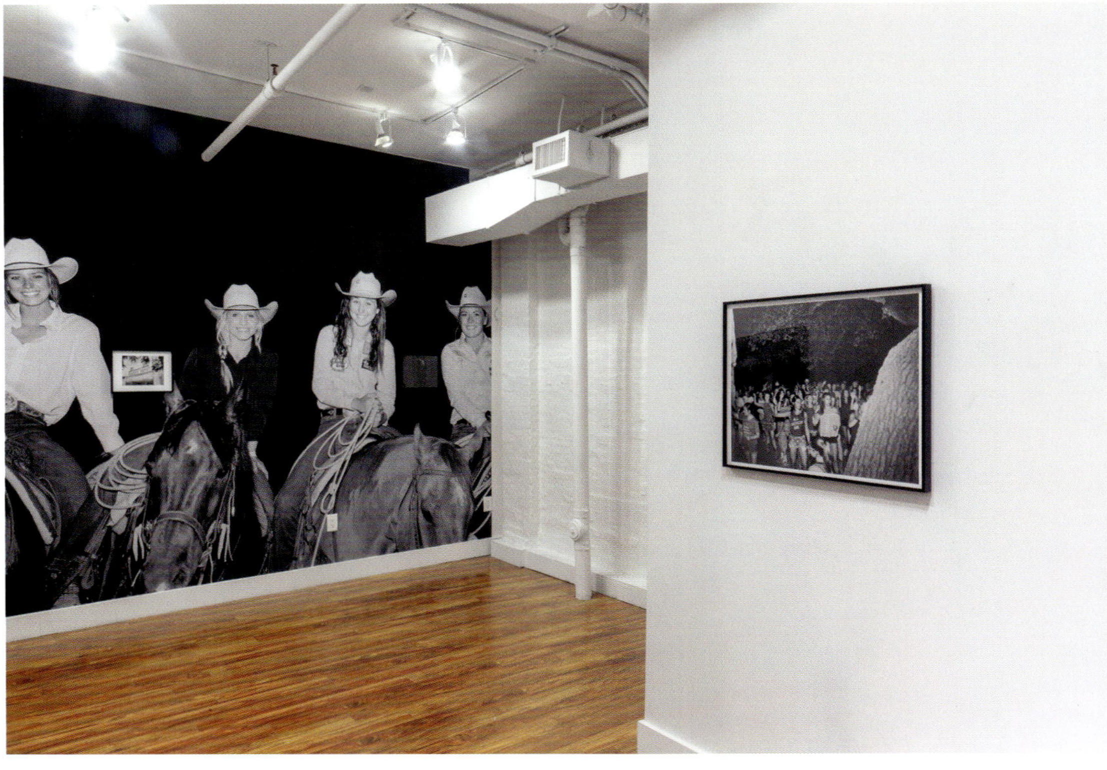

David Billet '17 &
Ian Kline '17
(Photography B.F.A.),
"Texas Is The Reason,"
Curated by Carl
Gunhouse,
The Java Project,
Brooklyn, NY, 2018

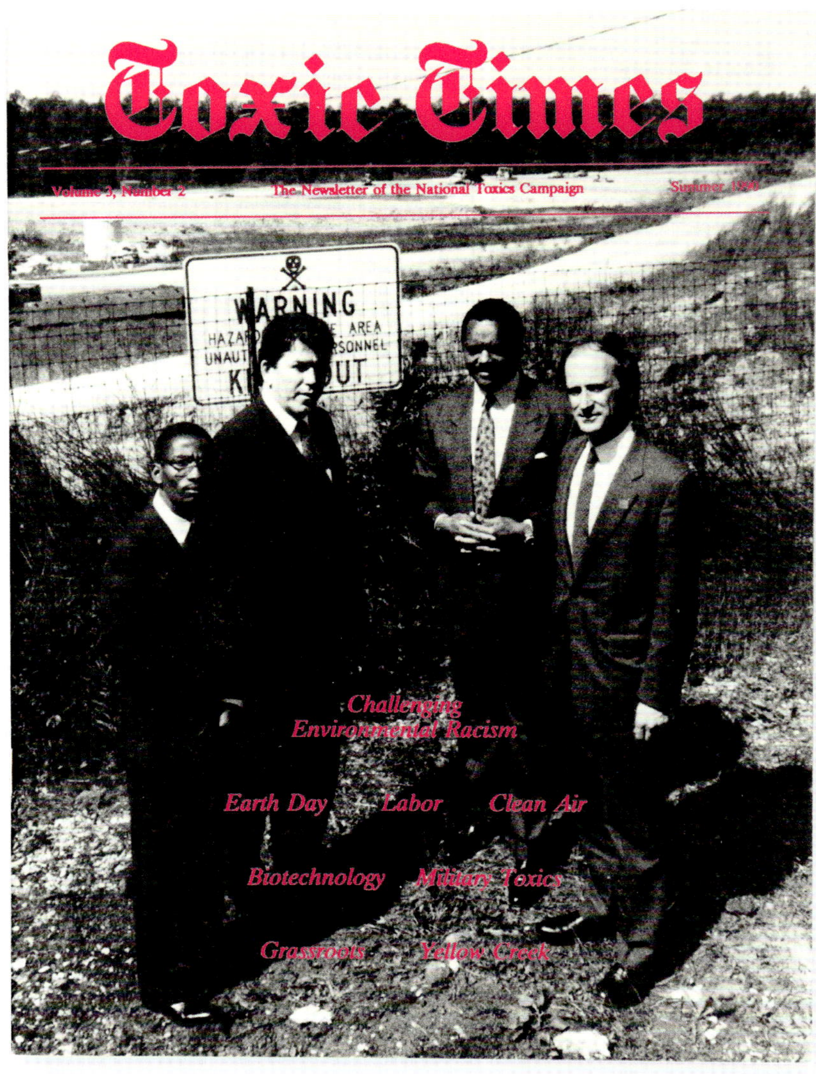

In the summer of 1978, Lois Gibbs, a resident of the upstate New York community of Love Canal, discovered that her child's elementary school was built on top of a chemical waste dump. Her kindergarten-age son had been suffering from frequent recurring illnesses, including liver disorders, epilepsy, asthma, and urinary issues. Mobilized by the suspicion that these circumstances were somehow interlinked, Gibbs formed the Love Canal Parents Movement. What began as a simple canvassing effort would later evolve into one of the nation's best-known grassroots environmental campaigns.

Nearly a decade later, activists formed the National Toxics Campaign (NTC), an organization committed to supporting local environmental causes, similar to what Gibbs and the Love Canal community had accomplished, on a national scale. The NTC began publishing a newsletter called *Toxic Times* to disseminate the efforts of the campaigns. As Michael Stein, the newsletter's former editor, said recently, these events were unfolding on a pre-Internet and predigital stage, so the newsletter was their primary means of sharing what was going on in local communities with a larger audience. "We were using super-old desktop-publishing software, the first version of Windows. It was painful and terrifying."

Stein and his colleagues knew that photography would be central in the campaign against corporate and government polluters. The cover of the summer 1990 issue of *Toxic Times* shows Louisiana state senator Cleo Fields, NTC executive director John O'Connor, Reverend Jesse Jackson, and Earth Day 1990 CEO Denis Hayes standing at a Superfund site in Baton Rouge, Louisiana, all gathered for the Southern Toxics Tour, which worked to educate low-income minority communities on hazardous materials.

A multigenerational and multiracial effort, *Toxic Times* also fostered a network of image sharing; organizations mailed in pictures from all over the country to be published in the newsletter. The exchange became a way to highlight the specificities of local experiences as symptomatic of a national crisis. "What else was there to prove what was going on?" Stein said. "We came together to teach people how to run an environmental campaign and help build that social movement." —**The Editors**